AF540319

EMERGING NANOTECHNOLOGY

ENCYCLOPAEDIA OF NANOTECHNOLOGY-III

EMERGING NANOTECHNOLOGY

By

Dr. M.P. Arora

Dept. of Zoology
M.M.H. Post Graduate College
Ghaziabad
(U.P.)

DISCOVERY PUBLISHING HOUSE PVT. LTD.
NEW DELHI-110 002

First Published-2008

ISBN 978-81-8356-279-9

Published by

DISCOVERY PUBLISHING HOUSE PVT. LTD.
4831/24, Ansari Road, Prahlad Street,
Darya Ganj, New Delhi-110002 (India)
Phone: 23279245 • Fax: 91-11-23253475
E-mail: dphbooks@rediffmail.com
dphtemp@indiatimes.com

Printed at:
Sachin Printers, Delhi

Preface

Everyone in the modern world knows what technology is. But what is nanotechnology? Taken from the Greek, *nano* means "one billionth part of" a whole. In modern parlance, it means very, very small. In today's world of smaller cell phones and portable computers, miniaturization is the current technology. Nanotechnology is the next step after miniaturization. In tomorrow's world, nanotechnology will be the new common technology. It will affect everyone of the planet and many change civilization as we know it.

Nanotechnology is the molecular engineering. It will soon create effective machines as small as DNA. It describes the ideas and techniques that are creating or new domain of science and technology. Nanotechnology holds the promise of revolutionizing field ranging from medical science to optical communications, but requires a substantial investment in time and money to make the promise become a readily. The results could be nothing less than a new industrial revolution.

In an undertaking as new as nanotechnology, many different approaches are studied in order to ascertain the best way to fashion working devices. At present, techniques can be divided into one of two general categories. One technique, termed the top-down method, physically manipulates molecules into nanostructures. The other technique, termed the bottom-up method, coaxes atoms and molecules to assemble themselves into nanostructures.

The present title **"Emerging Nanotechnology"** has been designed for undergraduate and post-graduate students as well as those involved in basic research in biological, biochemical and biophysical sciences.

In the preparation of this book large number of books and research papers have been consulted. So no authenticity is claimed.

The author expresses his gratitude to Mr. Wasan and staff of M/s Discovery Publishing House for their whole hearted co-operation in the publication of this book.

The author tried hard to be accurate and upto date in statement and realises the impossibility of completely avoiding errors therefore, the author will greatly appreciate having his attention called to any questionable statement.

Author

CONTENTS

1

INTRODUCTION

In 1346, when the Italian poet Franceso Petrarch scaled Mount Ventoux in southern France, he came down from the summit pondering the artificial limits dictated by his epoch's manner of thought. From this small seed bloomed a whole new insight and approach to life, sparking the explosion of art and science that came to be known as the *European Renaissance*. In the same spirit, a deep pondering of the limits of our own technological limits has given rise to explosive insights and ideas that could lead to nothing less than a new renaissance in science, reshaping fields ranging from communications to computing to medical science.

This renaissance is beginning to find its form in the Lilliputian world of *molecules* and *atoms*. The newly emerging field of nanotechnology builds and manipulates structures far too tiny for the unaided human eye to see, but the potential impact of its "*invisible*" machines is profound. And yet, like the revelations of Petrarch, a series of obstacles must be addressed and overcome before nanotechnology's full potential can unfold.

When atoms and molecules are manipulated to make structures called *nanomachines*, the physics that shape their behaviour radically changes. The puzzling and counterintuitive impact of quantum mechanics becomes magnified on the molecular and atomic level. In "The Art of Building Small," researchers George M. Whitesides and J. Christopher Love detail some of the hurdles standing in the path of fully functioning nanomachines.

Simply shrinking preexisting forms into *nanosize* and expecting them to work on a nanoscale is not a feasible approach. Forces and influences that have a nominal impact on the world we live in are major influences in the *nano world*. In "*Making Molecules into Motors*," some of the problems—and creative solutions—to this puzzle are explored.

Medical technology will most likely be the first beneficiary of the fruits of nanotechnology. In this field, simple structures can be made that are custom-designed to meet a very special purpose. "*Less Is More in Medicine*" explains how drug delivery; testing for the presence of specific genetic signatures, and even the reinforcing of artificial joints will probably be the first arenas to host nano technological innovations.

Microprocessors also stand to benefit in a revolutionary fashion from nanotechnology. As we rapidly approach the limit to silicon's usefulness as a microprocessor's raw material, alternative paths must be forged. "*The Incredible Shrinking Circuit*" details the myraid benefits of working with transistors 10 to 1,000 times smaller than the smallest scales attainable with silicon-based technology.

Revolution comes at a price, and that price is usually the abandonment of old and familiar manners of doing things. Today, as in Petrarch's day and age, the promise of a new approach must also contend with the entenched inertia of institutions quite content with the way things are. Yet, *nanotechnology* stands of the threshold of changing the very fabric of the world we know. And like the renaissance that swept through Europe so many centuries ago, nanotechnology will manifest itself in our future world in ways we can, at present, imagine, and most probably in ways that we could never dream.

Scrambling for funding is an essential aspect of any ground breaking science done today. Nanotechnology holds the promise of revolutionizing fields ranging from medical science to optical communications, but requires a substantial investment in time and money to make the promise become a reality. The result could be nothing less than a new industrial revolution.

Little Big Science

Albert Einstein, calculated the size of a single sugar molecule from experimental data on the diffusion of sugar in water. His work showed that each molecule measures about a nanometer in diameter. At a billionth of a meter, a nanometer is the essence of small. The width of 10 hydrogen atoms laid side by side, it is one thousandth the length of a typical bacterium, one millionth the size of a pinhead. one billionth the length of Michael Jordan's well-muscled legs. One nanometer is also precisely the dimension of a big windfall for research, Almost 100 years after Einstein's insight, the nanometer scale looms large on the research agenda. If Einstein were a graduate student today probing for a career path, a doctoral adviser would enjoin him to think small: "Nanotech, Albert, nanotech" would be the message conveyed.

After biomedical research and defense, *nanotechnology* has become the most highly energized discipline in science and technology. The field is a vast grab bag of stuff that has to do with creating tiny things that sometimes just happen to be useful. It borrows liberally from condensed-matter physics, engineering, molecular biology and large swaths of chemistry. Researchers who once called themselves materials scientists or organic chemists have transmuted into nanotechnologists.

Purist academic types might prefer to describe themselves as mesoscale engineers. But it's "*nano*" that generates the buzz. Probably not since Du Point coined its corporate slogan "*better things for better living through chemistry*" have scientists who engage in molecular manipulation so adeptly captured and held public attention this case, the votes of lawmakers in Washington who hold the research purse strings. "You need to come up with new, exciting, cutting-edge, at things in order to convince the budget- and policy-making apparatus to give you more money," remarks Duncan Moore, a former White House official who helped to organize the Clinton administration's funding push for nanotechnology.

With recognition has come lots of money—lots, that is, for something that isn't a missile shield. In USA, the National Nano technology Initiative (NNI), initiated by President Bill Clinton, is a multi-agency program intended to provide a big funding boost to *nanoscience* and *engineering*. The $422-million budget in the federal fiscal year 2000 marked a 56 percent jump in nano spending from a year earlier. The initiative is on track to be augmented for fiscal year 2002 by another 23 percent even while the Bush administration has proposed cuts to the funding programs of most of the federal agencies that support research and development. *Nanomania* flourishes eve In 2001 more than 30 nanotechnology research centers and interdisciplinary groups sprouted at universities; fewer than 10 existed in 1999. *Nanoism* does not, moreover, confine itself to the U.S. In other countries, total funding for nanotechnology jumped from $316 million in 1997 to about

$835 million last year, according to the National Science Foundation (NSF).

Interest in nano is also fueled, in an aberrant way, by the Visions of a fringe element of futurists who muse on biblical life spans, on unlimited wealth and, conversely, on a holocaust brought about by legions of uncontrollable self-replicating robots only slightly bigger than Einstein's sugar molecules.

When Clinton introduced the nanotechnology initiative he was long on vision and short on specifics: nanotech, he noted, might one day store the Library of Congress on a device the size of a sugar cube or produce materials with 10 times the strength of steel at a mere fraction of its weight. But this wasn't just the meanderings of a starry-eyed politician. Surprisingly, the science establishment itself is a little unclear about what it really means when it invokes nano. "It depends on whom you ask," Stanford biophysicist Steven M. Block told a National Institutes of Health symposium on nanotechnology last year in a talk that tried to define the subject. "Some folks apparently reserve the word to mean whatever it is they do as opposed to whatever it is anyone else does."

The definition is indeed slippery. Some of *nanotechnology* is not *nano*, dealing instead with structures on the micron scale (millionths of a meter), 1,000 times or more larger than a nanometer. Also, nanotechnology, in many cases, is not technology. Rather it involves basic research on structures having at least one dimension of about one to several hundred nanometers. (In that sense, Einstein was more a nanoscientist than a technologist.) To add still more confusion, some nanotechnology has been around for a while: nano-size carbon black particles (a.k.a. *high-tech soot*) have gone into tires for 100 years as a reinforcing additive, long before the prefix "*nano*" ever created a stir. For that matter, a vaccine, which often consists of one or more proteins with nanoscale dimensions, might also qualify.

But there is a there in both *nanoscience* and *nanotechnology*. The *nano-world* is a weird borderland between the realm of individual atoms and molecules (where quantum mechanics rules) and the macro world (where the bulk proper ties of materials emerge from the collective behaviour of trillions of atoms, whether that material is a steel beam or the cream filling in an Oreo). At the bottom end, in the region of one nanometer, *nano-land* bumps up against the basic building blocks of matter. As such, it defines the smallest natural structures and sets a hard limit to shrinkage: you just can't build things any smaller.

Nature has created *nanostructures* for *billennia*. But Mihail C. Roco, the NSF official who oversees the nanotechnology initiative, offers a more restrictive definition. The *emerging field*—new versus old nanotech—deals with materials and systems having these key properties: they have at least one dimension of about one to 100 nanometers, they are designed through processes that exhibit fundamental control over the physical and chemical attributes of molecular-scale structures, and they can be combined to form larger structures. The intense interest in using nanostructures stems from the idea that they may boast superior electrical, chemical, mechanical or optical properties—at least in theory.

Real-world nano, fitting Roco's definition, does exist. Sandwiching several nonmagnetic layers, one of which is less than a nanometer thick, between magnetic layers can produce sensors for disk drives with many times the sensitivity of previous devices, allowing more bits to be packed on the surface of each disk. Since they were first introduced in 1997, these giant magnetoresistive heads have served as an enabling technology for the multibillion-dollar storage industry.

New tools capable of imaging and manipulating single molecules or atoms have ushered in the new age of nano. The icons of this revolution are scanning probe microscopes—the scanning tunneling microscope and the atomic force micro scope, among others—capable of creating pictures of individual atoms or moving them from place to place. The IBM Zurich Research Laboratory has even mounted the sharp, nanometer-scale tips used in atomic force microscopes onto more than 1,000 microscopic cantilevers on a microchip. The tips in the Millipede device can write digital bits on a polymer sheet. The technique could lead to a data storage device that achieves 20 times or more the density of today's best disk drives.

Varied approaches to fabricating *nanostructures* have emerged in the nano world. Like sculptors, so-called *top-down practitioners* chisel out or add bulk material to a surface. Microchips, which now boast circuit lines of little more than 100 nano meters, are about to become the most notable example. In contrast, bottom-up manufacturers use self-assembly processes to put together larger structures—atoms or molecules that make ordered arrangements spontaneously, given the right conditions. Nanotubes—graphite cylinders with unusual electrical properties—are a good example of self-assembled nanostructures.

The dwindling size of circuits in electronic chips drives much of the interest in nano. Computer companies with large research laboratories, such as IBM and Hewlett-Packard, have substantial *nano programs*. Once conventional silicon electronics goes bust—probably sometime in the next 10 to 25 years—it's a good bet that new nano technological electronic devices will replace them. A likely wager, though not a sure one. No one knows whether manufacturing electronics using nano tubes or some other novel material will allow the relent less improvements in chip performance without a corresponding increase in cost that characterizes silicon chip making.

Even if molecular-scale transistors don't crunch zeroes and ones in the Pentium XXV the electronics fashioned by nanotechnologists may make their way into devices that reveal the secrets of the ultimate small machine: the biological cell. Bionano, in fact, is finding real applications before the advent of postsilicon nanocomputers. Relatively few nanotags made of a semiconductor material are needed to detect cellular activity, as opposed to the billions or trillions of transistors that must all work together to function in a nanocomputer. One company, Quantum Dot Corporation, has already emerged to exploit semiconductor quantum dots as labels in biological experiments, drug-discovery research, and diagnostic tests, among other applications.

Outside biology, the earliest wave of products involves using *nanoparticles* for improving basic material properties. For instance, *Nanophase Technologies*, one of the few companies in this field that are publicly traded, produces nano-size zinc oxide particles for use in sunscreen, making the usually white- coloured cream transparent because the tiny particles don't scatter visible light.

The government's nanotech initiative goes beyond sun screen. It envisages that nanostructured materials may help reduce the size, weight and power requirements of spacecraft, create green manufacturing processes that minimize the generation of unwanted by-products, and form the basis of molecularly engineered biodegradable pesticides. The field has such a broad scope—and basic research is still so new in some nanosubspecialties—that worries have arisen about its ability to deliver on ambitious technology goals that may take 20 years to achieve. "While nanotechnology may hold great promise, some scientists contend that the field's definition is too vague and that much of its '*hype*' may not match the reality of present scientific speculation," noted a Congressional Research Service report last year.

Any advanced research carries inherent risks. But *nanotechnology* bears a special burden. The field's bid for respectability is coloured by the association of the word with a cabal of futurists who foresee nano as a pathway to a techno-utopia: unparalleled prosperity, pollution-free industry, even something resembling eternal life.

In 1986—five years after IBM researchers Gerd Binnig and Heinrich Rohrer invented the scanning tunneling microscope, which garnered them the Nobel Prize—the book Engines of Creation, by K. Eric Drexler, created a sensation for its depiction of godlike control over matter. The book describes self-replicating nanomachines that could produce virtually any material good, while reversing global warming, curing disease and dramatically extending life spans. Scientists with tenured faculty positions and NSF grants ridiculed these visions, noting that their fundamental improbability made them an absurd projection of what the future holds.

But the visionary scent that has surrounded nanotechnology ever since may provide some unforeseen benefits. To many nonscientists, Drexler's projections for nanotechnology straddled the border between science and fiction in a compelling way. Talk of cell-repair machines that would eliminate aging as we know it and of home food-growing machines that could produce victuals without killing anything helped to create a fascination with the small that genuine scientists, consciously or not, would later use to draw attention to their work on more mundane but eminently more real projects. Certainly labeling a research proposal "nanotechnology" has a more alluring ring than calling it "applied mesoscale materials science."

Less directly, Drexler's work may actually draw people into science. His imaginings have inspired a rich vein of science-fiction literature. As a subgenre of science fiction—rather than a literal prediction of the future—books about Drexlerian nanotechnology may serve the same function as Star Trek does in stimulating a teenager's interest in space, a passion that sometimes leads to a career in aeronautics or astrophysics.

The danger comes when intelligent people take Drexler's predictions at face value. Drexlerian nanotechnology drew renewed publicity last year when a morose Bill Joy, the chief scientist of Sun Microsystems, worried in the magazine Wired about the implications of nanorobots that could multiply uncontrollably. A spreading mass of self robots—what Drexler has labeled "*gray goo*"—could pose enough of a threat to society, he mused, that we should consider stopping development of nanotechnology.

Among real chemists and materials scientists who haw now become nanotechnologists, Drexler's predictions have assumed a certain quaintness; science is nowhere near to being able to produce nanoscopic machines that can help revive frozen brains from suspended animation. Zyvex, a company started by a software magnate enticed by Drexlerian nanotechnology, has recognized how difficult it will be to create robots at the nanometer scale; the company is now dabbling with much larger micromechanical elements, which Drexler has disparaged in his books.

Even beyond meditations on gray goo, the nanotech field struggles for cohesion, Some of the research would have proceeded regardless of its label. Fusing "nano" and "technology" was an after-the-fact designation: IBM would have forged ahead in building giant magnetoresistive heads whether or not the research it was doing was labeled nanotechnology.

For the field to establish itself as a grand unifier of the applied sciences, it must demonstrate the usefulness of grouping widely disparate endeavors. Can scientists and engineers doing research on nano powders for sunscreens share a common set of interests with those working on DNA computing? In some cases, these crossover dreams may be justified. A semiconductor quantum dot originally developed for electronics and now being deployed to detect biological activity in cells is a compelling proof of principle for these types of trans-disciplinary endeavors.

If the nano concept holds together, it could, in fact, lay the groundwork for a new industrial revolution. But to succeed, it will need to discard not only fluff about nanorobots that bring cadavers back from a deep freeze but also the overheated rhetoric that can derail any big new funding effort. Most important, the basic nano science must be forthcoming to identify worth while nanotechnologies to pursue. Distinguishing between what's real and what's not in nano throughout this period of extended exploration will remain no small task.

2

GEOLOGICAL TIME SCALE

To understand the complete evolutionary changes in the organisms (plants and animals, one has to study the organisms of the distant past which are now extinct. The extinct forms of life, which are also the connecting links between the most primitive forms and recent forms, are preserved in the rocks as *fossils*. The study of these fossils constitutes the subject-matter of *Palaentology*.

A number of estimates have been proposed from time to time about the age of earth. For examples, *Archbishop Ussher* of Ireland in 1645 declared that the creation of earth took place 26th October 4004 B.C., at 9 o'clock in the morning. This was the adopted version of the *Bible* for a long time. *James Hutton* believed that geological time was unimaginably long. *Lord Kelvin* (1897), from a study of the flow of heat from the interior to the surface of earth, estimated the age of earth from 20-40 million years.

The most accurate method of determining the age of a rock is by radioactive decay studies. Certain radioactive elements like uranium and thorium, found in certain rocks, decay into lead. The rate of this radioactive disintegration is constant and in unaffected by external physical conditions like temperature, pressure, etc. The formula may be written as $U^{238} = Pb^{206} + 8He^{4} +$ heat. It has been determined that on gram of uranium gives on decay 1/7,600,000 gram of lead in one year. The lead: uranium ratio in a rock would thus give the age of the rock. The maximum age of the earth has thus been estimated to be around 5500 m.y. (million years). Other methods of dating, *e.g.*, potassium—argon method, radiocarbon method, etc. are also known today.

Geological Time Scale

By studying the fossils, present in the rocks, the geologists have divided the earth's past history into *Era* which are major divisions. The eras are divided into *periods*. The periods are subdivided into *Epoches*. These eras, periods and epoches are arranged on the time scale in an order of their age, and this arrangement is called, '*geological time scale*'. The first geological time scale was developed by *Giovanni Avduina*, an Italian Scientist, in 1760.

Eras

Based on the nature of the forms of life preserved as fossils in the rock, the earth's past history is divided into six eras.

I. Azoic
II. Archaeozoic or Eozoic (Primal life or dawn of life)
III. Proterozoic (Early life)
IV. Palaeozoic (Ancient life)
V. Mesozoic (Mediaeval life)
VI. Cenozoic (Recent life)

Periods

These are the fundamental units of the standard geologic time scale. There are 12 periods:

I. Periods of Palaeozoic era

There are 7 periods in Palaeozoic era:

1. Cambrian
2. Ordovician
3. Silurian
4. Devonian
5. Mississipian
6. Pennsylvanian
7. Permian.

II. Periods of Mesozoic era

There are three periods:

8. Triassic
9. Jurassic
10. Cretaceous.

III. Periods of Cenozoic era

There are two periods:

11. Tertiary
12. Quaternary.

Epoches

There are seven epoches. The epoches of cenozoic's period are:

I. Quaternary period

There are two epoches:

1. Recent
2. Pleistocene.

II. Tertiary period

It is subdivided into five epoches:

3. Pliocene
4. Miocene
5. Oligocene
6. Eocene
7. Palaeocene.

Azoic Era

This was the time, in which life was not present on the earth. During this time, earth was formed, cooled and underwent many changes. These changes made favourable conditions for the appearance of living organisms. The rock of this time are igneous type without any fossil.

Archaeozoic Era

The manner in which life originated is not known. It is probable that sometime in early Archaeozoic Eras the earliest living matter that could reproduce itself got generated under very favourable conditions of temperature, humidity, light, etc., by the combination of certain chemicals. This first living matter may have been very similar to protoplasm. Filterable viruses which represent an intermediate stage between living and non-living entities, may have been the first to appear on the surface of the globe in the early Archaeozoic Period. Unicellular Protozoa and Protophyta may have been also present.

According to *Stirton* (1959), the oldest known fossils are blue-green algae, fungi, and probably a flagellate described from Ontario, Canada. The age of the rock in which these algae occur has been estimated to be about 2000 m.y. (million years) by radiometric methods.

Proterozoic Era

It may be called the '*Age of Primitive Invertebrates*'. Life was in the form of algae, precipitations, siliceous sponge spicules, remains of

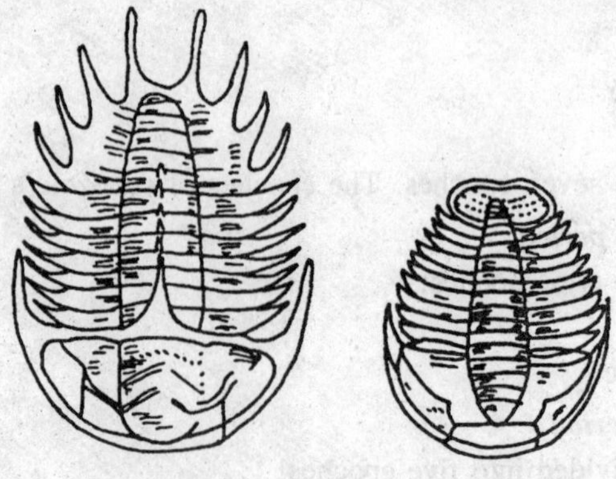

Fig. 2.1.

radiolarians, impressions of jelly fishes, a simple brachiopod (*Lingullela montana*) as well as deposits of graphite. These first organisms were quite possibly soft bodied. Reducing bacteria must also have been predominant during this era because the chief iron-ore deposits of the world are found in the Proterozoic era.

The fossil record from the rocks is much less probably due to absence of hard parts in the organisms to be preserved as fossils. The fossils which were formed must have been destroyed due to effect of pressure and high temperatures.

This era is estimated to have lasted 3.5 billion years.

Palaeozoic Era

This fossil record in this era are very extensive. Almost all the major invertebrate phyla are represented even in the early Palaeozoic era. The fossils of vertebrates also present late in the era because the early chordates were soft-bodied and could not be preserved as fossils.

Palaeozoic era is divided into following periods:

Cambrian Period

The name Cambrian was proposed by *A. Sedgwick* in 1835. It is taken from *Cambria* the Latin name of City of Wales.

This period started by the melting of glaciers, slow rise of sea level and warm climate.

Almost all the main invertebrate phyla had existed during this period as also the main phyla of plants. Trilobite were the most dominant forms of life, the important genera being *Olenellus*, *Redlichia* in the early Cambrian times; *Paradoxides* in the middle Cambrian and *Crepicephalus* in late Cambrian. Small primitive brachiopods and

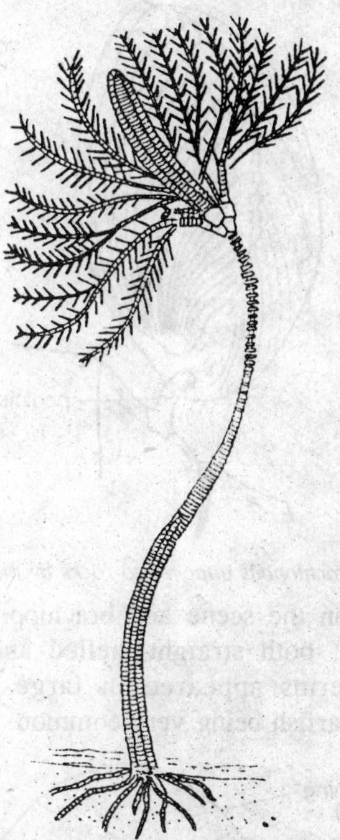

Fig. 2.2. Crinoids or sea-lilies which look like flowers growing in the ocean.

gastropods with simple shells are found in early Cambrian rocks. Some foraminifers and radiolarians have also been found. A few graptolites are recorded in the Cambrian rocks. Echinoderms have been recorded from Cambrian rocks of Australia while some bryozoans from Russia and Canada. The first land plants also appear in this period.

Ordovician Period

The name Ordovician was proposed by *Charles Lapworth* in 1879, on the basis of the ordovices, an ancient tribe, that formerly inhibited the City of Wales.

During this period, conditions became more favourable as the climate was uniformly warm with glacial activities in certain regions.

This was the "*Age of Giant Mollusca*". Graptolites, which started in late Cambrian, show their maximum development during Ordovician. They in fact serve as index fossils in Ordovician rocks. Trilobites were still abundant and reached the climax of their evolution. First

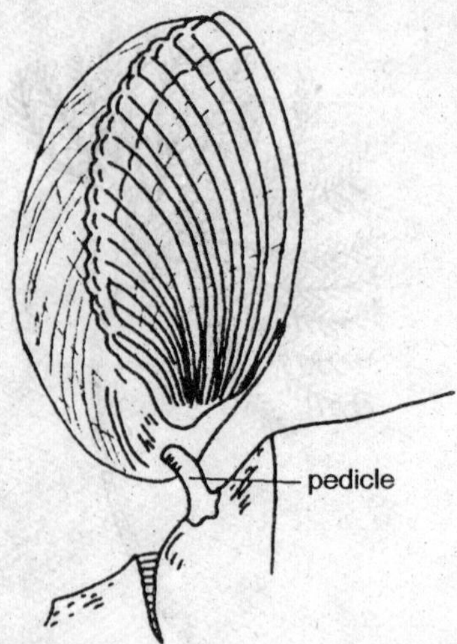

Fig. 2.3. Brachiopods attached to rocks by their pedicles.

ostracods appeared on the scene and brachiopods developed harder shells. Cephalopods, both straight-shelled and coiled type, were abundant. Echinoderms appeared in large numbers and show diversification, the starfish being very common.

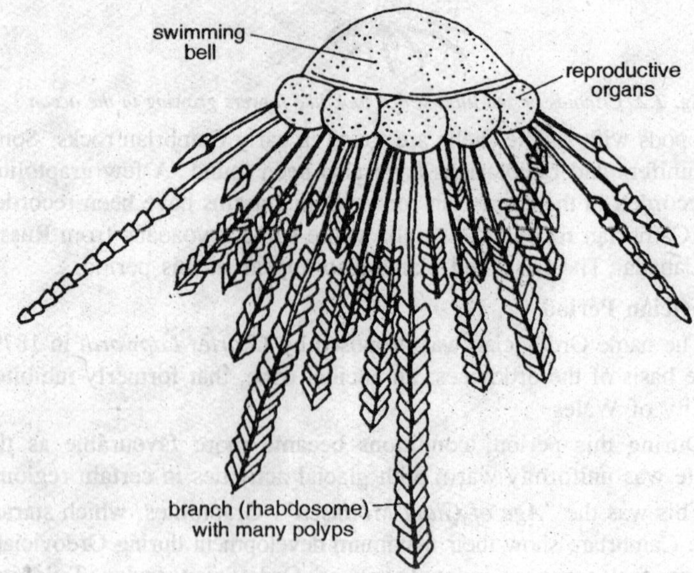

Fig. 2.4. Graptolite colony.

Ordovician is also characterised by first appearance of conodonts which are very important marine environment indicators. Land plants are meagre. First appearance of fish-like animals Ostracoderms is also recorded in Ordovician rocks. This Period probably marks the first appearance of vertebrates.

Silurian Period

The term Silurian was proposed by *Murchison* in 1835 to designate the rocks present on the borders of Wales and England. This area was inhibited by Silures, an ancient tribe, inhibited Wales.

The climatic conditions were mild and some areas were arid. Land elevation at places caused formation of deep continental seas. Rise of *Caledonian Mountains* in British Isles, Scandinavia and N. Greenland and in N. Africa and East Central Asia.

The first air-breathing invertebrates on the earth were scorpions which appeared in the Silurian Period. Trilobites now started waning. Bryozoans were extremely abundant. Brachiopods continued their development and first *Spirifer* appeared in Silurian Period. Graptolites also started a decline at the end of Silurian Period.

The fish is found to have developed lower jaws in the Silurian Period. The organisation of the Silurian fishes is similar in many respects to that of the fishes of today.

Although algae were still dominant but some terrestrial primitive ferns also evolved.

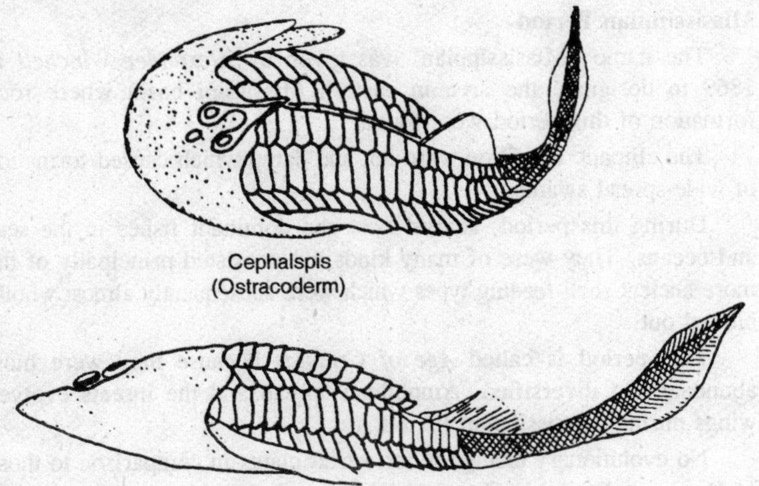

Fig. 2.5. An ostracoderm, Cephalaspis.

Devonian Period

The name Devonian was proposed by *Sedgwick* and *Murchison* in 1839 after Devonshire, a country of South-Western England, where rock formation of this period were first found.

The land elevation continued. Consequently larger seas were further divided into smaller seas. Due to fall of temperature, climate remained cold any dry. This period was marked by great volcanic activities and the formation of coal, oil and gas.

During the period, brachiopod developed wonderfully. Corals and bryozans were also abundant.

The Devonian Period is called the '*Age of Fishes*' because of their great number. Plate-skinned fishes, sharks and bony fishes were common. The Devonian ancestors of modern lung fish, Choanichthyes, were able to survive dry period by burrowing in the moist beds of streams or lakes. The lobed-fin fish, crossopterygians were characterized by the presence of lobed rounded bases of the paired fins. These were on the direct line of evolution from fish to terrestrial vertebrates. They became extinct at the closer of Palaeozoic except *Latimeria* and *Melania*.

From the late Devonian Rocks of Pennsylvania, the foot-print named, *Thinopus antiquas* gave the evidence of first amphibian life.

Tall terrestrial Psilophytes, club-mosses and horse-tails occur in the rock of Scotland. First wingless insect also appeared in this period.

Mississippian Period

The name `Mississippian' was given by *Alexander Winchell* in 1869 to designate the area in eastern Mississipi basin where rock formation of this period were found.

The climatic condition were hot and damp which caused formation of wide-spread swamps.

During this period, sharks were the dominant fishes in the seas and oceans. They were of many kinds but consisted principally of the more ancient shell-feeding types which were subsequently almost wholly blotted out.

This period is called *Age of Crinoids* because they were most abundant and diversified. Amphibian spread and the insects evolved wings during Mississippian Period.

No evolutionary change in terrestrial plants in comparison to those of Devonian Period is observed in this period.

Pennsylvanian Period

Term 'Pennsylvanian' was proposed by *H.S. Williams* in 1819 which refers to the state Pennsylvania.

The climatic conditions were disturbed so that the forests burried under the swamps and turned, in due course of time, into present-day coal mines. For this reason, Mississippian and Pennsylvanian Periods are often combined together under a head, *Carboniferous Period* (i.e., coal bearing).

Amongst animals, insects developed abundently both in size and diversity. Cockroaches reached such a large number that sometimes this period is called the '*Age of Cockroaches*'. Dragon flies were also abundant. Among vertebrates, amphibians showed maximum diversity and specialization during this period. True sharks and sea-likes were common. Towards the close of this period appeared the first animal, *Tuditanus punctulatus*, showing the reptilian affinities, from Stegocephalians.

Among plants, ferns dominated and bryophytes also appeared.

Permian Period

The term 'Permian' was given by *Murchison* in 1841, because fossil record of the period were found in Russian province of Perm.

The cold and dry condition caused widespread glaciation. Extensive land elevation caused the formation of high mountains. Consequently, shallow seas were emptied and their bottom became exposed as salty deserts. Melting of glaciers gave rise to rivers upon the land.

Great diversification in reptiles took place and first mammal like reptiles appeared in the Permian times. Many important groups of marine invertebrates now dwindled. Permian marks the complete extinction of Trilobites which were very important and dominant during the early Palaeozoic. Due to the cold climates, the insect developed larval adaptations seen in the metamorphosis of present day insects.

Among terrestrial animals, spiders, scorpions, centipedes, snails etc. were present.

Among plants swampy trees declined and were replaced by tall gymnosperms. The true conifers and cycads became most abundant and trees somewhat similar to date palms appeared.

Mesozoic Era

In contrast to the dominance of marine animals throughout Palaeozoic Era, the Mesozoic Era is characterized by the dominance of land animals, the most spectacular of which were giant dinosaurs which ruled the earth for thousands of years. That is why this era is known as '*Age of Reptiles*'. This Era is divided into three Periods.

Triassic Period

This name was give by *F. Von Alberti*.

This period was characterized by hot and dry climate and land elevation which helped in expression of deserts and continents.

Snails, bivalved molluscs, insects and sea-urchins diversified and spread. This period is characterized by the diversification of reptiles. Dinosaurs spread upon land; Ichthyosaurs (fish-like) in water; lizard-like Plesiosaurs in water; and bird like Pterosaus in air. The egg-laying mammals also made their appearance.

During this period horstails, clubmosses and seed fern declined, Gymnosperms progressed fast.

Fig. 2.6. Brontosaurus.

Jurassic Period

The name was given by *A. Von Humboldt* in 1799. This Period has been named after the Jura Mountains of Switzerland and France. The climates during this Period were mild throughout the world.

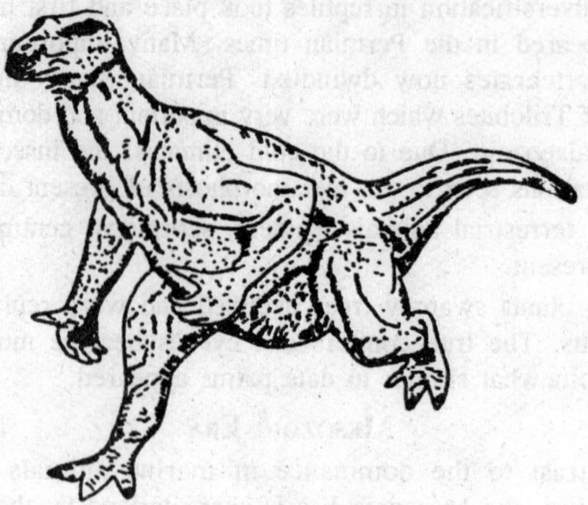

Fig. 2.7. Iguanodon.

Consequently both invertebrate and vertebrate fauna thrived. Among the invertebrates, belemnites and ammonites were most abundant; the latter, in fact, served as a basis for division of Jurassic rocks into eighty *stages* and thirty-three *zones* by Oppel.

Jurassic is the age of giant dinosaurs like Ichthyosaurs which were more than 50 feet long, and whose skull along measured more than 8 feet. The first bird *Archaeopteryx*, which is the connecting link between reptiles and birds has been found in the Solnhofen limestone of Jurassic Period. Similarly the first mammals are recorded in Jurassic Period.

Among plants gymnosperms continued spreading. Advanced seed ferns gave rise to the first Angiosperms which were dicotyledonous.

Crataceous Period

The term was given by *J.J. d'omalus* d'Halloy in 1882 derived from Latin word *Creta*=chalk.

The climate was warm and somewhat uniform over the most earth surface, but became cooler later on. The seas spread widely over the continents in the earliest past of this period. The Larmide Orogeny changed the surface relief of the earth which affected the climates. The changed climate accelerated the evolution of mammals, birds, flowering plants etc. while caused complete extinction of certain groups of animals like the dinosaurs, ammonites and rudistids.

In the sea, sharks and bony fishes flourished. Snakes first appeared in this period. New varieties of birds appeared. Mammals were still small and less numerous. Marsupials diverged from the archaic forms. Placental mammals appeared.

Fig. 2.8. Stegosaurus.

During this period the dinosaurs became extinct because of their heavy body and small brain, lack of adaptability, coldbloodness, herbivorous food etc.

Among plants, the gymnosperms declined. Dicot angiosperms flourished and tall oak, maple etc. formed thick forests. The first monocots appeared.

Cenozoic Era

It is known as "*Age of mammals*". It was marked by great adaptive radiation in birds, insects and flowering plants. Wide variety of animals got food and protection as a result of development and spread of deciduous trees, herbs and grasses.

The cenozoic era is divided into two periods, and each period into epochs.

Tertiary Period

This period includes five apoches.

Palaeocene epoch

In the beginning the climate was fairly warm but gradually it became colder.

After complete extinction of dinosaurs, nine new orders of terrestrial mammals evolved. Among these were the first primitive carnivores, lemurs, tarsiers etc. Modern groups of birds evolved. Plants also underwent a corresponding evolutionary change.

Eocene epoch

Climate remained less hot. These was a widespread erosion of mountains.

All the modern orders of mammals were present. The first horse *Eohippus* or Hyracotherium evolved also first camels which were originally of the size of a dog or even smaller evolved during this period. The horse had progressed to the *Epihippus* stage by this time. Pigs, rats, monkeys, whales and seals appeared.

Angiosperms spread and evolved, giving rise to several modern types. Heavy forest and limited grass land characterized in this period.

Oligocene epoch

Climate remained hot. The level of land decreased.

Most of the families of mammals were recognizable by this time. Larger brontotheres reached their climax in early Oligocene Period and suddenly died out before middle Oligocene-Period. The Oligocene camel *Poebrotherium* was of the size of a small sheep. The true flesh-eaters like dogs, cats, sabertoothed cats and other small animals, as well as rodents and insect-eating animals became most abundant. The horse had reached the *Miohippus* stage. It was of the size of a sheep with three toes in each foot. Tortoises of medium-size were abundant. Hawks and vultures hovered in the sky and the 'American pigs' on the ground. Near the end of the Oligocene Period the first New World monkeys appeared.

Among plants tropical and subtropical forests, flowering plants and monocots became widespread.

Miocene epoch

In Miocene epoch, the Alps reached their hights and Himalayas were pushed up. So the climate remained somewhat cold. New mountain ranges such as Cascade, Sierra etc., formed upon America continent.

This Period saw the rise and rapid evolution of grazing mammals, long slender-limbed camels, *Oxydactylus*, bear dogs, small foxes, American pigs etc. were particularly abundant. The horse had reached *Parahippus* stage, a major step in evolution towards modern horse. A free interchange of old and new world terrestrial mammals took place among which cats, dogs, horses and mastodonts were especially common. In Late Miocene Period, the horse had reached *Merychippus* stage of evolution in North America. In Eurasia, the Miocene Epoch is marked by the appearance of hyenas, deer, giraffes, and true antelopes.

Terrestrial plants were more progressive.

Pliocene epoch

This period started with subsidence of land in many places and ended with the beginning of one or two Ice Ages. The climate became still cooler.

The mammals became more specialized and reached their peak of evolution. Camels, horses, rhinoceroses, etc. were abundant during this period. Camels had grown larger in size, and the horse had evolved into several genera. *Pliohippus*, which ultimately gave rise to modern horse, evolved during this period from its Miocene ancestors. *Hipparion* fauna was typical of Pliocene Period in Eurasia and has distributed from Western Europe and North Africa to South-western Asia, India, Burma and China. Characters found in the fossil jaws and teeth of *Dryopithecus*, found in Siwalik rocks of India, showed affinities with chimpanzee, gorilla and modern man. The lack of a land bridge between North and South America in early Pliocene Period inhibited the free migration and dispersal.

Thick forests declined due to cold. Tail and woody plants were replaced by soft monocot plants.

Quaternary Period

This period is divided into two epoches:

Pleistocene epoch

There are evidences in four *Ice Ages* alternating with warmer intervals. Widespread glaciation during this epoch lowered the sea level. Land connection between European continent and Britain, and Siberia and Alaska were established. Due to these land connections dispersal of horses, camels, deer, cats, saber, dogs, rodents etc., took place.

The modern horse, *Equus* originated.

The first evidence of a primitive man ***Homo heidelbergensis***, recognised by a massive mandible, was found in the middle Pleistocene rocks of Heidelberg, Germany. The Peking man, *Pithecanthropus pekinensis*, lived in the mildy cooler climates in parts of China and adjoining areas. The ape man of Java, *Pithecanthropus erectus*, found among Pleistocene fauna of Java, attests to the evidence of land bridge between Java an mainland of Asia. The entire Pleisotocene Period shows dominance of man.

Recent epoch

The climate has again become hot. Supremacy of man as the wisest and dominant species has been established. With increasing mental development, gradually he got control over water, air and land. He is trying to get control over plants other than earth. Thus we can say that this epoch is the '*Age of man*'.

3

Live Molecules Outside the Solar System

Is there life on Mars? The question is an interesting and legitimate scientific one, quite unrelated to the fact that generations of science-fiction writers have populated Mars with creatures of their imagination. Of all the extraterrestrial bodies in the solar system Mars is the one most like the earth, and it is by far the most plausible habitat for extraterrestrial life in the solar system. For that reason major objective of the Viking mission to Mars was to search for evidences of life.

The two Viking spacecraft were launched from Cape Canaveral in the summer of 1975. Each spacecraft consisted of an orbiter and an attached lander. When the spacecraft arrived at Mars in July and August of 1976, each was put in a predetermined orbit around the planet, and the search for a landing place began. Cameras aboard the orbiters were the principal source of information on which the choice of the landing sites was based; important data also came from infrared sensors on the orbiters and from radar observatories on the earth. The sole consideration in the final selection of the sites was the safety of the spacecraft. It would be a mistake to suppose, however, that the sites were therefore without biological interest. Biological criteria dominated the initial decisions as to the latitude at which each spacecraft would land. Once the latitudes had been chosen there was relatively little difference between sites at different longitudes.

On command from the earth each lander separated from its orbiter. With the help of its retroengines and parachute it dropped to the surface of Mars. Both orbiters continued to circle the planet, operating their own scientific instruments and relaying to the earth data transmitted from the landers. Both landings were in the northern hemisphere of Mars, and the Martian season was summer. (Mars has seasons like those on the earth, but each season lasts approximately twice as long. The Martian year is 687 Martian days; each Martian day, named a sol by the Viking team to distinguish it from a terrestrial day, is 24 hours 39 minutes long). On July 20, 1976, the *Viking I* lander came to rest in the Chryse Planitia region of Mars, some 23 degrees north of the equator, Six weeks later the *Viking* 2 lander settled down in the Utopia Planitia region, some 48 degrees north of the equator. In longitude the two landers are separated by almost exactly 180 degrees, thus placing them on opposite sides of the planet. Since the instrumentation of the two landers is identical, the difference in their landing sites is the only distinction between them.

The first biologically significant task carried out by each lander was the analysis of the Martian atmosphere. Life is based on the chemistry of light elements, notably carbon, hydrogen, oxygen and nitrogen. To be suitable as an abode of life a planet must have those elements in its atmosphere. Spectroscopic observations from the earth and from spacecraft that had flown past Mars in previous years had already shown that carbon dioxide was the principal component of the Martian atmosphere. Small quantities of carbon monoxide, oxygen and water vapour had also been detected. Nitrogen had not been detected in any form, however, and atmospheric theory suggested that Mars had lost most of its nitrogen in the past.

Each Viking lander analyzed the atmosphere by means of two mass spectrometers. One spectrometer, operating during the descent to the surface, sampled and analyzed the atmospheric gases every five seconds. The second spectrometer operated on the ground. The results showed that the atmosphere near the ground was approximately 95 per cent carbon dioxide. 2.5 per cent nitrogen and 1.5 per cent argon, and that it also held traces of oxygen, carbon monoxide, neon, krypton and xenon. At both landing sites the atmospheric pressure was 7.5 millibars. (The atmospheric pressure at sea level on the earth is 1.013 millibars).

Since the Viking spacecraft revealed that nitrogen is indeed present in the Martian atmosphere, we can say that the elements necessary for life are available on Mars. Missing from the list of gases, however is one critically important compound: water vapour. Although earlier measurements had shown that traces of water vapour are present in the Martian atmosphere, the quantity varies with season and place.

The Viking orbiters carried out a survey of water vapour over the entire planet with infrared spectrometers. The results showed that the highest concentration of atmospheric water vapour was at the edge of the north polar cap (the summertime hemisphere), and that the concentration fell off toward the south (the opposite of what is found on the earth). In the polar region the amount of water vapour in the atmosphere would form a film only a tenth of a millimeter thick if all of it were to be condensed on the planet's surface. At the landing sites the concentration of water vapour ranged between 10 and 30 per cent of the concentration at the pole.

These numbers put into quantitative terms a long-known fact about Mars: It is a very dry place. Mars has ice at its poles, but nowhere on its surface are there oceans or lakes or any other bodies of liquid water. The absence of liquid water is related to the dryness of the atmosphere through a fundamental law of physical chemistry: the phase rule. The phase rule states that for liquid water to exist on the surface of a planet the pressure of the water vapour in the atmosphere must at some times and in some places be at least 6.1 millibars. The Viking measurements imply that the vapour pressure of water at the surface of Mars in the northern hemisphere is at most .05 millibar, even if all the water vapour is concentrated in the lower atmosphere. At that low pressure liquid water cannot remain in the liquid phase: depending on the temperature, it must either freeze or evaporate. By the same token raindrops cannot form in the Martian atmosphere and ice cannot melt on the Martian surface.

The extreme dryness presents a difficult problem for any Martian biology. Liquid water is essential for life on the earth. All terrestrial species have high and apparently irreducible requirements for water; none could live on Mars. If there is life on Mars, it must operate on a different principle as far as water is concerned. If Mars had a more favourable environment in the past, however, and if the planet did not dry up too fast, species may have had time to evolve and adapt to present conditions. Pictures made by the *Mariner 9* spacecraft, which went into orbit around Mars in 1971, suggested that Mars may indeed have had running water on its surface in the past. The pictures from the Viking orbiters have confirmed that impression. The evidence consists of channels in the Martian desert that resemble dry riverbeds. There seems to be little doubt that the channels were carved by rapidly flowing liquid, and there is widespread agreement that the most probable liquid is water.

If liquid water once existed on Mars, could life have arisen on the planet? If the life evolved to meet changing conditions, could it exist there still? There is no way to settle these questions by deductive reasoning or even by experimentation in laboratories on the earth. They can be answered only by the direct exploration of Mars, and that is what the Viking spacecraft did.

Five different types of instrument on each Viking lander were involved in the search for evidences of life: two cameras for photographing the landscape, a combined gas chromatograph and mass spectrometer for analyzing the surface for organic material and three instruments designed to detect the metabolic activities of any microorganisms that might be present in the soil engineers and managers whose joint efforts made all the Viking projects possible. They work in universities, industrial laboratories and the National Aeronautics and Space Administration and its field centers. Their names are recorded in the growing technical literature dealing with this historic mission.

Each of the Viking landers carried two cameras of the facsimile type, which built up a picture of the scene by scanning it in a series of narrow strips. Such cameras make pictures slowly, but they are rugged and versatile. Their resolution was moderately high: a few millimeters at a distance of 1.5 meters. They produced pictures in black and white, in colour and in stereo. The two cameras on each lander could between them survey the entire horizon around the spacecraft.

As life-seeking tools cameras have inherent advantages and disadvantages. Their chief advantage lies in the fact that a picture contains a large amount of information. In principle it would be possible to prove unequivocally the existence of life on Mars with a single photograph. For example, if a line of trees were visible on the horizon or if foot-prints appeared on the ground in front of the spacecraft one morning, there would be no room for doubt that there is life on Mars. Another advantage lies in the fact that pictorial evidence is independent of all assumptions about the chemistry and physiology of Martian organisms. The organisms need not respond in certain ways to certain substances or treatments in order to be recognized. The cameras could identify, say, a mushroom made of titanium as a form of life if one were to sprout up from under a rock in the course of the mission. Of course, reliance on pictorial evidence rests on its own set of assumptions about the morphology of living things. The most obvious disadvantage of the camera as a life-seeking instrument is the fact that an entire world of life can exist below the camera's limit of resolution.

Of all the results of the Viking mission the wonderful photographs of the Martian desert at the two landing sites are the most impressive. The photographs have been eagerly scanned by alert and hopeful eyes, but no investigator has yet seen anything suggesting a living form. The next step was to analyze the soil for any organic constituents. Among the elements carbon is unique in the number, variety and complexity of the compounds it can form.

The special properties of carbon that enable it to form large and complex molecules arise from the basic structure of the carbon atom. The structure enables the carbon atom to form four strong bonds with other atoms, including other carbon atoms. The molecules thus formed are very stable at ordinary temperatures, so stable, in fact that there seems to be no limit to the size they can attain. The connection

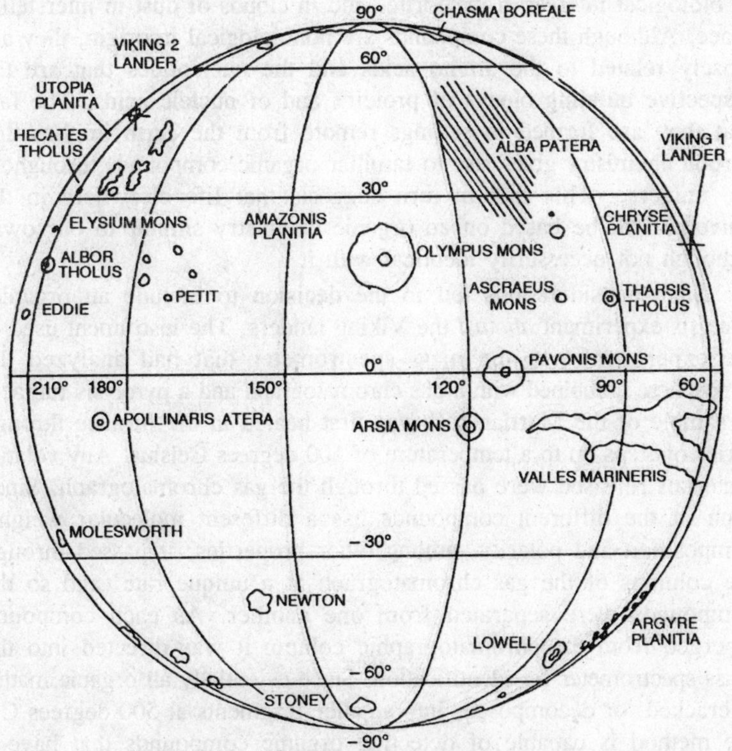

Fig. 3.1. Locations of the Two Viking Landers are indicated on this map of Mars, which shows some of the major geological features of the planet.

between life and organic chemistry (that is, the chemistry of carbon) rests on the fact that the attributes by which we identify living things—their capacity to replicate themselves, to repair themselves, to evolve and to adapt—originate in properties that are unique to large organic molecules. It is the highly complex information-rich proteins and nucleic acids that endow all the living things we know, even "simple" ones such as bacteria and viruses, with their essential nature. No other element, including that favourite of science-fiction writers silicon has the capacity carbon has to form large and complex structures that are so stable. It is no accident that even though silicon is far more abundant than carbon on the earth, it has only minor and nonessential roles in biochemistry. Biochemistry is largely a chemistry of carbon.

Such fundamental facts lead to the conclusion that wherever life arises in the universe it will most likely be based on carbon chemistry. That view has been strengthened by the discovery of organic compounds of biological interest in meteorites and in clouds of dust in interstellar space. Although these compounds are nonbiological in origin, they are closely related to the amino acids and the nucleotides that are the respective building blocks of proteins and of nucleic acids. The fact that they are formed in settings remote from the earth implies that carbon chemistry gives rise to familiar organic compounds throughout the universe. This fact in turn suggests that life elsewhere in the universe will be based on an organic chemistry similar to our own, although not necessarily identical with it.

Such considerations led to the decision to include an organic-analysis experiment *aboard* the Viking landers. The instrument used in the experiment was the mass spectrometer that had analyzed the atmosphere combined with a gas chromatograph and a pyrolysis furnace. A sample of the Martian soil was first heated in the furnace through series of steps up to a temperature of 500 degrees Celsius. Any volatile materials released were passed through the gas chromatograph. Since each of the different compounds has a different molecular weight, composition and polarity, among other properties, it passed through the columns of the gas chromatograph at a unique rate, and so the compounds were separated from one another. As each compound emerged from the chromatographic column it was directed into the mass spectrometer for identification. Since essentially all organic matter is cracked, or decomposed, into smaller fragments at 500 degrees C., the method is capable of detecting organic compounds that have a wide range of molecular weights.

Two soil samples were analyzed at each landing site. The only organic compounds detected were traces of cleaning solvents known to have been present in the apparatus. The fact that the solvents were detected shows the instruments were functioning properly. The heated samples gave off carbon dioxide and a small amount of water vapour; nothing else was found.

This result is surprising and weighs heavily against the existence of biological processes on Mars. The combined gas chromatograph and mass spectrometer aboard each Viking lander is a sensitive instrument, capable of detecting organic compounds at a concentration of a few parts per billion, a level that is between 100 and 1,000 times below their concentration in desert soils on the earth. Even if there is no life on Mars, it has been supposed the fall of meteorites onto the Martian surface would have brought enough organic matter to the planet to have been detected. Because Mars is near the asteroid belt, from which meteorites originate, it is believed to receive a much larger number of meteorite impacts than either the earth or the moon. Indeed, a question that was frequently discussed before the Viking spacecraft were launched was whether or not it would be possible to distinguish biological organic matter on Mars from the meteoritic organic matter that was expected to be present. The absence of organic matter at the parts-per-billion level, however, suggests that on Mars organic compounds are actively destroyed, probably by the strong ultraviolet radiation from the sun.

The other experiments aboard the Viking landers searched not just for organic matter in the soil but for living organisms. On the earth microorganisms such as bacteria, yeasts and molds are the hardiest of all species. There are few places on the earth where microbial forms do not live; they are the last survivors in environments of extreme temperature and aridity. The reasons for their hardiness are interesting but need not detain us here. Suffice it to say that if there is life on Mars, the chance of detecting it would be maximized by searching for microorganisms in the Martian soil.

Each Viking lander carried three instruments designed to detect the metabolic activities of soil microorganism. First, the gas-exchange experiment was designed to detect changes in the composition of the atmosphere caused by microbial metabolism. Second, the labeled-release experiment was designed to detect decomposition of organic compounds by soil microbes when they were fed with a nutrient. Third, the pyrolytic-release experiment was designed to detect the synthesis of

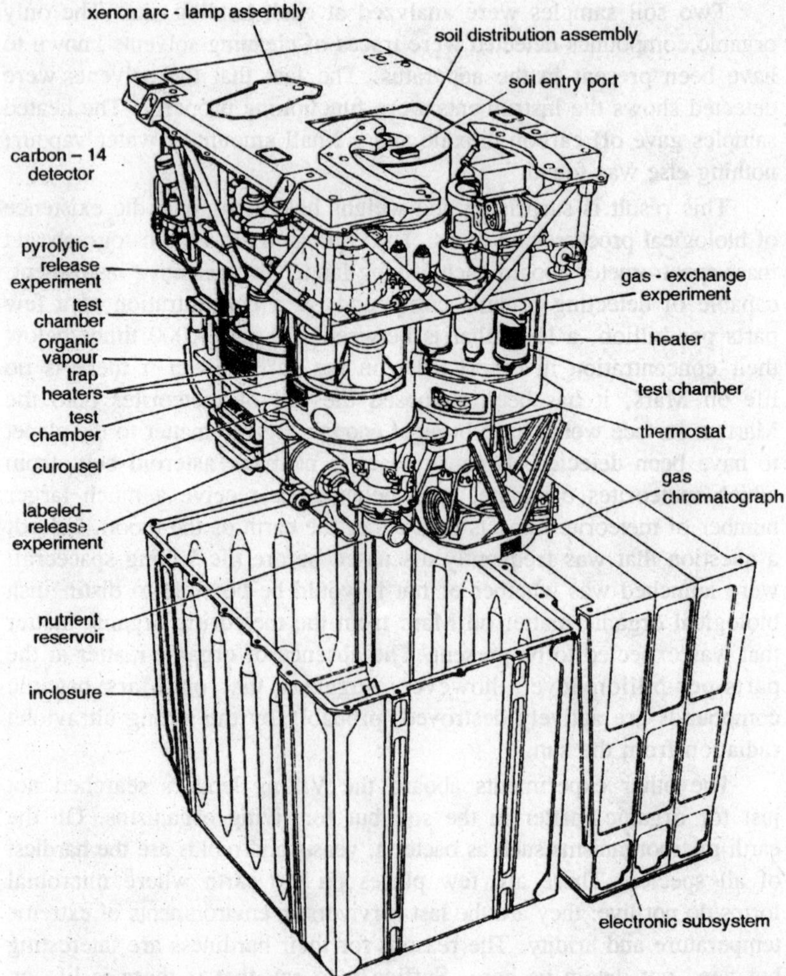

Fig. 3.2. Biological laboratory aboard both Viking spacecraft occupies a volume of only one cubic foot.

organic matter in Martian soil from gases in the atmosphere by either photosynthetic or non-photosynthetic processes.

All three experiments analyzed portions of each sample of Martian soil. All the experiments detected chemical changes of one kind or another in the soil. All the experiments are now completed, and some of the changes they observed suggest biological processes. There has been much discussion both within the team of Viking investigators and outside it as to the best way to interpret the findings. Are the changes due to biological responses or are they just chemical reactions we would like to believe are biological? Indeed, since life is a form of chemistry, how can the two be told apart?

One way to decide whether or not a process is biological is to test its sensitivity to heat. Living structures are highly organized and fragile, and they are destroyed by temperatures that leave many chemical reactions unaffected. A process that is insensitive to heat is thus likely to be a nonliving chemical reaction, but a process that is sensitive to it could be either living or nonliving.

The decision as to whether a heat-sensitive process is biological or not must be based on additional evidence. In the end, however, the judgement is based on Occam's razor: the traditional principle that the hypothesis most likely to be correct is the one that accounts for the maximum number of observations with the minimum number of assumptions.

The gas-exchange experiment and the labeled-release experiment were frankly terrestrial in orientation. In both experiments a nutrient medium composed of an aqueous solution of organic compounds was mixed with a sample of Martian soil. Since liquid water cannot exist on Mars, the experiments could not be conducted under Martian conditions; the test chambers had to be heated to prevent the water from freezing and pressurized to prevent it from boiling. Both experiments were based on the universal property of terrestrial organisms to evolve gas as they metabolize food. If a sample of soil from the earth is moistened with a nutrient solution, the microorganisms in the soil take up the nutrients and convert them partly into more microorganisms (that is, the population of microorganisms grows) and partly into various by-products, including gases. Among the gases given off in microbial metabolism are carbon dioxide, methane, nitrogen, hydrogen and hydrogen sulfide. On the earth gases evolved by one species of organisms are eventually consumed by other species of organisms. In that way the light elements at the earth's surface are continually cycled through the biosphere and the atmosphere.

In the gas-exchange experiment a complex nutrient solution was added to a sample of Martian soil in a closed chamber, and the gases were analyzed periodically by means of a gas chromatograph. The experiment proceeded in two stages. In the first stage a small volume of the nutrient solution was introduced into the soil chamber in such a way that it humidified the chamber without actually wetting the soil, and the resulting gases were analyzed several times. In the second stage a large volume of the nutrient was poured into the chamber, saturating the soil. With the soil now in direct contact with the medium the main part of the experiment began. The soil was incubated for nearly seven months, so that whatever microorganisms might be in the sample had enough time to signal their presence by producing or consuming gases. During the period of incubation the atmosphere in the chamber was periodically analyzed.

The findings of the first stage of the experiment were both surprising and simple. Immediately after the soil sample was humidified carbon dioxide and oxygen were rapidly released. The release of the gases ceased soon after it had begun but not before the pressure in the chamber had risen measurably. At the Chryse site in a period of little more than one sol the quantity of carbon dioxide in the incubation chamber of the *Viking I* lander increased by a factor of five and the quantity of oxygen increased by a factor of 200. At the Utopia site the increases were less, but they were still considerable.

The rapidity and the brevity of the response recorded by both landers clearly suggested that the process observed was a chemical reaction not a biological one. The appearance of the carbon dioxide is

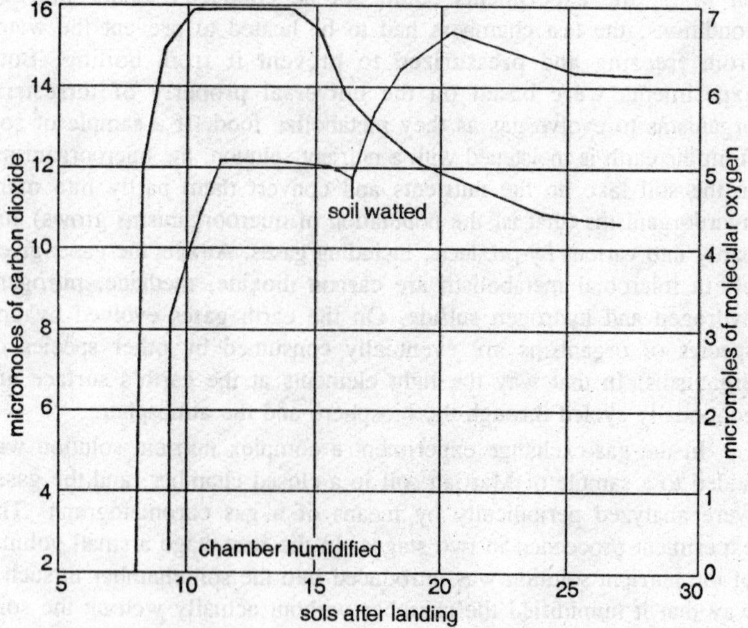

Fig. 3.3. Results of the Gas-exchange experiment, according to data of Vance I.

readily explained. Carbon dioxide gas would be expected to be adsorbed on the surface of the dry Martian soil; if the soil was exposed to a very humid atmosphere, the gas would be displaced by water vapour. The appearance of the oxygen is more complex. The production of so much oxygen seems to require an oxygen-generating chemical reaction, not just a physical liberation of pre-existing gas. It is likely that the oxygen was released when the water vapour decomposed an oxygen-rich compound such as a peroxide. Peroxides are known to decompose if they are exposed to water in the presence of iron compounds, and according to the X-ray fluorescence spectrometer aboard each Viking lander the Martian soil is 13 per cent iron.

At both landing sites the second phase of the gas-exchange experiment was anticlimactic. When the soil sample was saturated with the nutrient medium and incubated carbon dioxide continued to be released. The production of the carbon dioxide gradually tapered off, however and the oxygen gradually disappeared. The slow increase in the amount of carbon dioxide was probably a continuation of the reaction in the humid stage of the experiment. The disappearance of the oxygen also can be easily explained: one of the ingredients of the nutrient medium was ascorbic acid, which combines readily with oxygen. And so after seven months it became clear that everything of interest had happened in the humid stage of the experiment, before the soil came in contact with nutrient! What the gas-exchange experiment detected was not metabolism but the chemical interaction of the Martian surface material with water vapour at a pressure that has not been reached on Mars for many millions of years.

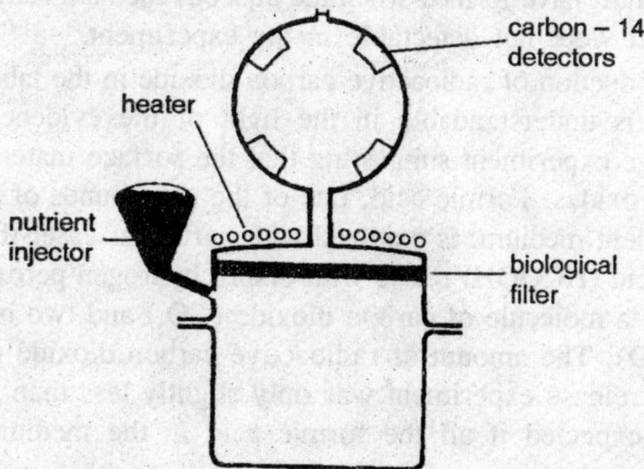

Fig. 3.4. Labeled-release experiment on the Viking landers tested the Martian soil for microorganisms that could metabolize simple organic, or carbon, compounds.

The labeled-release experiment differed from the gas-exchange experiment in several ways. The nutrient medium employed was a simpler one containing only a few cosmically abundant organic compounds such as formic acid (HCOOH) and the amino acid glycine (NH_2CH_2COOH). All the compounds were labeled with atoms of the radioactive isotope carbon 14. The labeled-release instrument was designed to detect radioactive gases, principally carbon dioxide, released when the nutrient medium was added to a sample of soil. The number of radioactive disintegrations in gases can be counted quite efficiently, so that the labeled-release experiment is faster and more sensitive than the gas-exchange experiment in detecting microbial activity in terrestrial soil. The labeled-release experiment's sequence of operations did not include a humid stage as such, but it attempted to accomplish the same end by injecting a volume of nutrient medium that was insufficient to wet the entire soil sample but sufficient to humidify the chamber. If the experiment worked on Mars as planned, subsequent injections of the medium, which were controlled by commands sent from the earth, brought the medium into contact with some soil that had previously been wetted and with other soil that had been humidified but not wetted.

As in the gas-exchange experiment, immediately after the nutrient medium was added to the soil in the labeled-release experiment, gas surged into the chamber. The release of gas tapered off soon after the first sol. The gas, undoubtedly carbon dioxide, was radioactive, showing that it had been formed from the radioactive compounds of the medium and not from compounds in the Martian soil. Non-radioactive gases, which also must have formed when the aqueous medium came in contact with the soil were not detectable in the experiment.

The production of radioactive carbon dioxide in the labeled-release experiment is understandable in the light of the evidence from the gas-exchange experiment suggesting that the surface material of Mars contains peroxides. Formic acid, one of the compounds of the labeled-release nutrient medium, is oxidized with particular ease: if a molecule of formic acid (HCOOH) reacts with one of hydrogen peroxide (H_2O_2), it will form a molecule of carbon dioxide (CO_2) and two molecules of water ($2H_2O$). The amount of radioactive carbon dioxide given off in the labeled release experiment was only slightly less than what would have been expected if all the formic acid in the medium had been oxidized in this way.

If the source of the oxygen released in the humid stage of the gas-exchange experiment was indeed peroxides in the soil decomposed

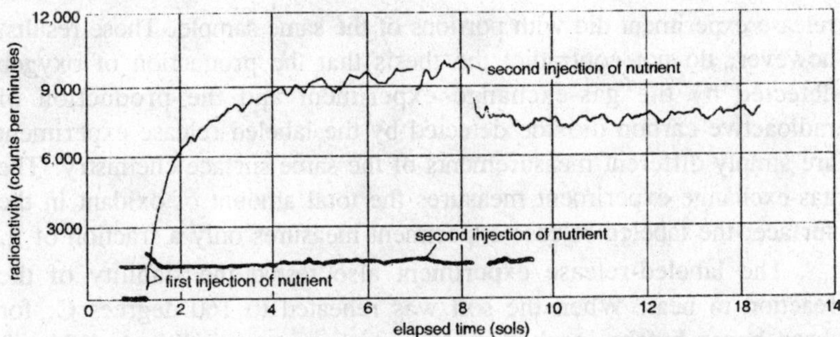

Fig. 3.5. Result of the Labeled-release experiment for the first sample of soil analyze at the Chryse site were indeed consistent with the result of the gas-exchange experiment.

by water vapour, then in the labeled-release experiment all the peroxides should also have been decomposed by the first injection of nutrient. Thus the next injection should have evolved no addi-tional radioactive gas inspite of the fact that part of the sample presumably had not yet been wetted by the medium. That proved to be the case. When a second volume of medium was injected into the chamber, the amount of the gas in the chamber was not increased; indeed, it decreased. The decrease is explained by the fact that carbon dioxide is quite soluble in water; when fresh nutrient medium was added to the chamber, it absorbed some of the carbon dioxide in the head space above the sample.

This result was obtained with all the samples tested by the labeled-release experiment at both Viking sites. In that respect the results of the labeled-release experiment did not parallel the results of the gas-exchange experiment. At both sites both experiments tested soil gathered from the ground's exposed surface; at the Utopia site the experiments also tested soil gathered from under a rock. Although the labeled-release experiment found essentially no difference in the amount of gas released by any of the samples, the gas-exchange experiment recorded about three-fourths as much carbon dioxide from the surface samples at the Utopia site as it had from the surface samples at the Chryse site, and it recorded even less carbon dioxide from the sample from under the rock.

The gas-exchange experiment also recorded less oxygen from the samples from the Utopia region, but the interference of the ascorbic acid in the complex nutrient medium of that experiment makes it difficult to quantify the difference. In every case, however the gas-exchange experiment detected considerably more gas than the labeled-release experiment did with portions of the same sample. Those results, however, do not contradict the thesis that the production of oxygen detected by the gas-exchange experiment and the production of radioactive carbon dioxide detected by the labeled-release experiment are simply different measurements of the same surface chemistry. The gas-exchange experiment measures the total amount of oxidant in the surface; the labeled-release experiment measures only a fraction of it.

The labeled-release experiment also tested the stability of the reaction to heat. When the soil was reheated to 160 degrees C. for three hours before incubation, the reaction was abolished. When it was heated to 46 degrees for the same length of time, the magnitude of the reaction was reduced by about half. These results have been regarded by some as evidence in favour of the hypothesis that the reaction is biological. The results are of course consistent with such a hypothesis, but they are also consistent with a chemical oxidation in which the oxidizing agent is destroyed or evaporated at relatively low temperatures. A variety of both inorganic peroxides and organic peroxides could probably have produced the same results.

The third microbiological experiment, the pyrolytic-release experiment, differed from the gas-exchange and labeled-release experiments in two respects. First, it attempted to measure the synthesis of organic matter from atmospheric gases rather than its decomposition. Second, it was designed to operate under the conditions of pressure, temperature and atmospheric composition that actually obtain on Mars, since those are the conditions under which any form of Martian life must exist. In practice the conditions in the chamber were a reasonably good approximation of Martian conditions except for the temperature, which stayed warmer than the outside temperature because of heat sources within the spacecraft.

A sample of Martian soil was sealed in a chamber along with some Martian atmosphere. A quartz window in the chamber admitted simulated Martian sunlight from a xenon arc lamp. Into this Martian microcosm small amounts of radioactive carbon dioxide and radioactive carbon monoxide were introduced. Both gases are present in the Martian atmosphere but not in radioactive form. After five days the lamp was turned off, the atmosphere was removed from the chamber and the soil was analyzed for the presence of radioactive organic matter.

First the soil was heated in the pyrolysis furnace to a temperature high enough to crack any organic compounds into small volatile fragments. The fragments were swept out of the chamber by a stream of helium and passed through column that was designed to trap organic molecules but allow carbon dioxide and carbon monoxide to pass through. The radioactive organic molecules were thus transferred from the soil to the column and at the same time were separated from any remaining gases of the incubation atmosphere. The organic molecules were released from the column by raising the column's temperature. Simultaneously the radioactive organic molecules were decomposed into radioactive carbon dioxide by copper oxide in the column. The carbon dioxide was then carried by the stream of helium into a radiation counter. If organic compounds had been synthesized in the soil, they would be detected as radioactive carbon dioxide; if no organic compounds had been synthesized, no radioactive carbon dioxide would have been formed.

Surprisingly, seven of the nine pyrolytic-release tests executed on Mars gave positive results. The two negative results were obtained at

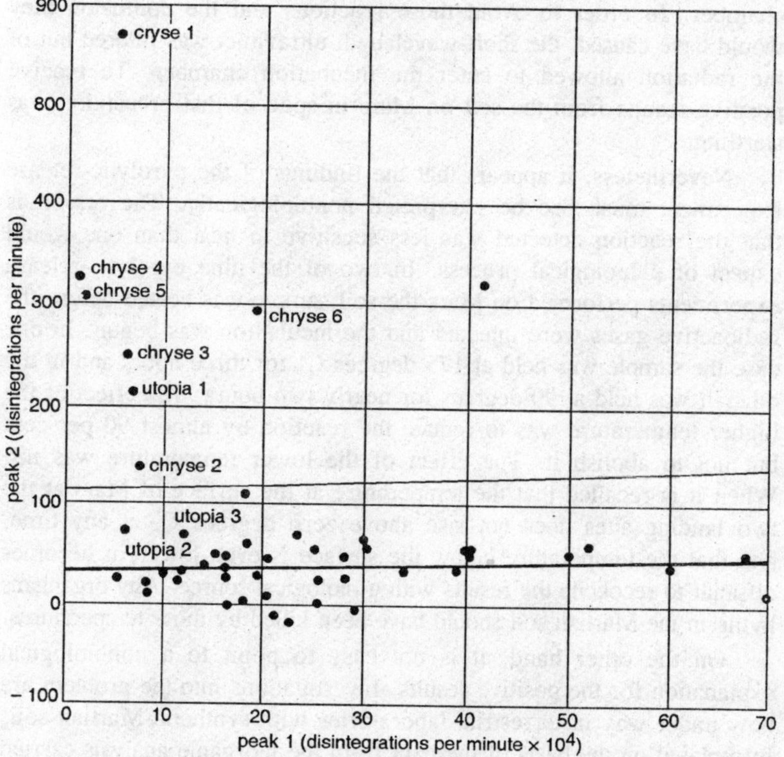

Fig. 3.6. Results of Pyrolytic-release experiment are shown for all the samples tested on Mars.

the Utopia site, but a third sample tested at Utopia was positive. This third sample was actually incubated in the dark, implying that light may not be required for the reaction. The amount of carbon fixed in the soil by the experiment was small: enough to furnish organic matter for between 100 and 1,000 bacterial cells. The quantity is so small, in fact, that it could not have been detected by the organic analysis experiment. The quantity is nonetheless significant; it was surprising that in such a strongly oxidizing environment even a small amount of organic material could be fixed in the soil.

Even more significant, the pyrolytic-release instrument had been rigorously designed to eliminate nonbiological sources of organic compounds. During the development of the experiment it had been found that in the presence of short-wavelength ultraviolet radiation, carbon monoxide spontaneously combined with water vapour to form organic molecules on glass, quartz and soil surfaces in the experimental chamber. In order to avoid those reactions and the confusion they would have caused, the short-wavelength ultraviolet was filtered out of the radiation allowed to enter the incubation chamber. To receive positive results from the soil on Mars in spite of that precaution was startling.

Nevertheless, it appears that the findings of the pyrolytic-release experiment must also be interpreted nonbiologically. The reason is that the reaction detected was less sensitive to heat than one would expect of a biological process. In two of the nine pyrolytic-release experiments performed on Mars the soil sample was heated before the radioactive gases were injected and the incubation was begun. In one case the sample was held at 175 degrees C. for three hours and in the other it was held at 90 degrees for nearly two hours. The effect of the higher temperature was to reduce the reaction by almost 90 per cent but not to abolish it. The effect of the lower temperature was nil. When it is recalled that the temperature at the surface of Mars at the two landing sites does not rise above zero degrees C. at any time, and that the temperature below the surface is even lower, it becomes difficult to reconcile the results with a biological source. Any organisms living in the Martian soil should have been killed by those temperatures.

On the other hand, it is not easy to point to a nonbiological explanation for the positive results. Investigations into the problem are now under way in terrestrial laboratories with synthetic Martian soils formulated on the basis of the data from the inorganic analysis carried out by the Viking landers. The solution to the puzzle will probably also explain why the organic-analysis experiment detected no organic material in the Martian surface. Until the mystery of the results from the pyrolytic-release experiment is solved, a biological explanation will continue to be a remote possibility.

Even though some ambiguities remain, there is little doubt about the meaning of the observations of the Viking landers: At least those areas on Mars examined by the two spacecraft are not habitats of life. Possibly the same conclusion applied to the entire planet, but that is an intricate problem that cannot yet be addressed. The most surprising finding of the life-seeking experiments is the extraordinary chemical reactivity of the Martian soil: its oxidizing capacity, its lack of organic matter down to the level of several parts per billion and its capacity to fix atmospheric carbon (presumably into organic molecules) at a still lower level. It seems Mars has a photochemically activated surface that, due to the low temperature and the absence of water, is maintained in a state far from chemical equilibrium.

These conclusions drawn from the results of the life-seeking experiments on the Viking landers are undeniably disappointing. The discovery of life would have been much more interesting, to say the least. There are doubtless some who, unwilling to accept the notion of a lifeless Mars, will maintain that the interpretation I have given is unproved. They are right. It is impossible to prove that any of the reactions detected by the Viking instruments were not biological in origin. It is equally impossible to prove from any result of the Viking experiments that the rocks seen at the landing sites are not living organisms that happen to look like rocks. Once one abandons Occam's razor the field is open to every fantasy. Centuries of human experience warn us, however, that such an approach is not the way to discover the truth.

Life Outside the Solar System

The evolution of stars and the evolution of living organisms appear to be completely dissimilar processes. The differences between the two can be explained in terms of how the particles of matter are bound together and how energy is exchanged among them. Indeed, the evolution of living organisms represents one outcome of stellar evolution. It was the steady flow of energy from the sun for four or five billion years that brought about the biological developments on earth, culminating in the emergence of intelligent organisms able to contemplate the whole remarkable story. If the sun had a different history, life would not have appeared in its immediate vicinity.

Astronomical evidence acquired in recent years indicates that what has happened here is probably not unique. Two decades ago it was thought that the solar system might have originated in a near-collision of the sun and another star, which event supposedly pulled away enough matter from the sun to from the planets. Because such encounters must be rare events, they would give rise to few stars with a company of planets. Today most astronomers believe that a star is formed by the condensation of a cloud of dust and gas. This hypothesis much more readily explains the origin of the solar system and is supported by the observation that more than half of the stars in our galaxy are double or multiple systems. In fact, it now appears that most stars are accompanied by other stars or by planets, though the latter must be so small as to escape sure detection by the present instruments of astronomy. Thus the appearance of life—even the appearance of mind—may be far from unusual events in the universe.

On the other hand, certain critical conditions must be satisfied if life processes are to be initiated and maintained. In the first place the star must shine long enough and steadily enough to permit life to evolve. The star must also be hot enough to warm up a habitable zone deep enough to offer a reasonable chance that a planetary orbit will fall within it. And the planet must ply a stable orbit within this zone. The number of stars that have given rise to life must therefore be considerably smaller than the immediately suggested by the dust-cloud hypothesis.

Taking account of all these factors, what is the probability that man will ever be able to visit the life-bearing planet of another star, or that the earth will receive a visitor from such a planet? Does the possibility justify taking measures now-in advance of a visit—to put existing technology to the task of scanning the sky for signals from intelligent organisms outside the solar system, or for transmitting signals in the hope they may be heard? If so, what sort of signais should be listened for or sent?

A reliable answer to these questions calls for a somewhat closer consideration of the conditions critical for life and an estimate of how often they are likely to be satisfied in the evolution of stars. The first condition is a steady and prolonged flow of energy. How long it takes life to evolve may be judged from the single instance available: Here on earth rational animals evolved from inanimate matter in about three billion years. Biological evolution proceeds by the purely random process of mutation, that is, the unpredictable occurrence of novel chemical processes among the fantastically numerous reactions that constitute life. Since the process is a random one, the laws of probability suggest that the time-scale of evolution on earth should resemble the average time-scale for the development of higher forms of life anywhere. The rate at which mutations occur is of course a variable in the calculation. It is affected by the electromagnetic and corpuscular radiation from the parent star, and the amount of such radiation that reaches the surface of any planet is governed in turn by the magnetic field of the planet, the depth and composition of its atmosphere and other factors. But since mutation itself is a random process, the introduction of a few more variables does not affect the calculation greatly. Furthermore, a higher mutation rate does not necessarily accelerate evolution, because most mutations are harmful. The more frequently they occur, the greater the chance that an individual will suffer an injurious mutation. In order to favour natural selection, mutations should be rare, perhaps as rare as in the evolution of life on earth.

It is somewhat easier to estimate the time-scale of stellar evolution. In contrast to the random nature of biological evolution, stellar evolution is governed by the universal law of gravitation and by a relatively small number of thermonuclear reactions. When a star begins to form in a cloud of dust and gas, gravitational attraction among the gas and dust particles causes the cloud to condense until the pressure raises the temperature within it to the point at which the thermonuclear reactions that convert hydrogen to helium begin. The tremendous quantities of energy liberated by these reactions now set up a counterpressure from the center of the star that exactly balances the force of gravitational contraction. In this state of equilibrium the star shines for a much longer time than that required for its condensation out of dust and gas.

The vast majority of the visible stars are in this phase of their evolution. They are called "main-sequence" stars because their luminosity (energy output per unit of time) plotted on graph against their surface temperature places them in sequence in a narrow band. What the chart shows is that the hottest stars are also the most luminous. Luminosity depends upon mass, and so the point at which a star appears on the main sequence depends primarily upon the mass of material incorporated in it during condensation. The hottest stars are designated by the letter O, followed in descending order by stars classified B, A, F, G, K and M; these classifications are usually called spectral type. The adjectives "early" and "late," which have nothing to do with the age of the star, are often used before the spectral-type designation to denote further relative temperature differences. An early F-type star is hotter than a late one, which in turn has a higher temperature than an early G-type star such as our sun. The classification is further refined by a number from 0 to 9 written after the letter. Thus the class of early B-type stars includes those from B0 to B4; and the late B-type stars, those from B5 to B9.

The length of time a star remains in the equilibrium state on the main sequence can be calculated from the total mass of hydrogen in its core and the rate at which the hydrogen is consumed. Both factors in the calculation are determined by the position of the star on the main sequence. Luminosity, which is an index of the rate of fuel consumption, increases as the fourth power of the star's mass. The most massive stars thus use up their substance most rapidly and so have the shortest lifetimes in the equilibrium state. In general a star will evolve away from the main sequence when the core in which the hydrogen has been consumed has a mass of about 12 per cent of that of the entire star. With the exhaustion of fuel in the central furnace, gravitational contraction takes over again, heating up the interior until the thermonuclear reaction spreads to outer layers. The star now leaves the main sequence and in a comparatively brief period of time evolves into a red giant or supergiant. Its evolution thereafter cannot presently be predicted in detail. But somehow, perhaps through the rapid loss of mass by ejection, it ends up as a hot, faint, dense object known as a white dwarf. A majority of the intrinsic variable stars, of novae and of nova-like objects are apparently in the stage between red giants and white dwarfs.

In leaving the main sequence a star releases so much energy that it would destroy life on any of its planets. Thus the main-sequence stage of the star's evolution is the only important one so far as life is concerned. The O and early B stars are the most massive, but since they burn much more rapidly than the smaller, less luminous stars, their life on the main sequence lasts only a million to 10 million years. The small M stars, in contrast, remain on the main sequence for more than 100 billion years. None of the early stars in the O, B and A groups has a stable lifetime longer than three billion years; they cannot therefore sustain biological evolution long enough for intelligent organisms to appear on any of their planets. Closer calculation shows that only the stars farther down the sequence than F4 maintain their equilibrium for a sufficient length of time to bring biological evolution to its culmination.

If time were the only critical condition, then the late-K and M stars would stand the greatest chance of having life-bearing planets. But these stars have low luminosity, and the habitable zones in which planets might travel about them must be quite narrow. A more luminous star can obviously warm up a larger space than a less luminous one. It is like a fire in a field on a cold night: the bigger the fire, the wider the zone around it in which the temperature will be neither too hot nor too cold. Thus the chance of finding a planet and intelligent life within the habitable zone of any particular M or even late-K star is quite small. However, since there are 10 times as many M stars as G stars, the total number of life-bearing planets traveling around the M stars may not be negligibly small.

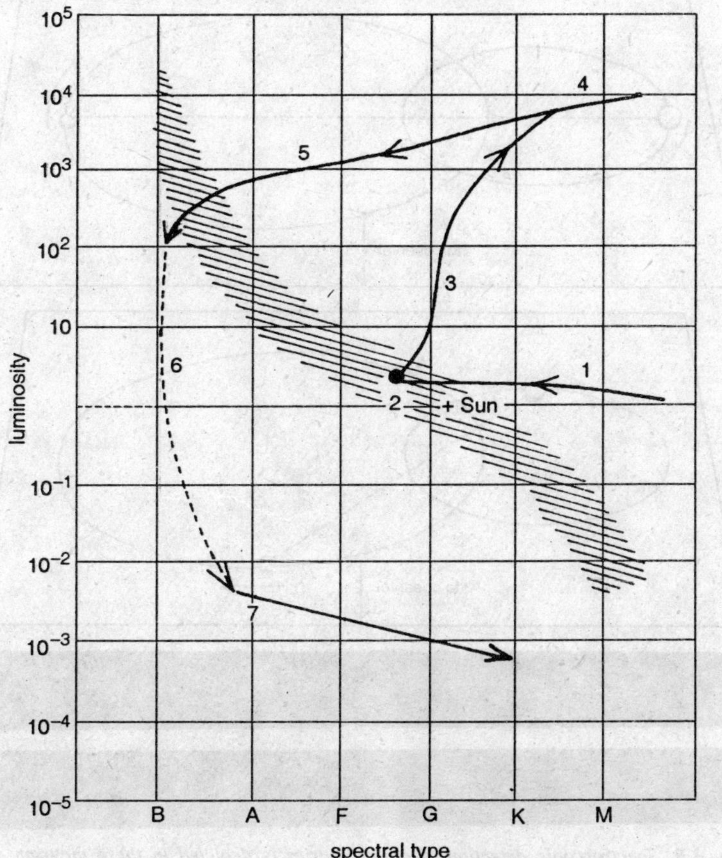

Fig. 3.7. Evolutionary path of a star with a mass 1.2 times that of the sun is traced.

On the basis of their lifetimes and the depth of their habitable zones, stars of the late-F, G and early-K types seem to offer the most favourable environments for life. Approximately 10 per cent of stars fall in these types. Out of the 200 billion stars in the Milky Way, therefore, some 20 billion might foster intelligent life.

This number is reduced considerably upon consideration of the third critical condition; the maintenance of stable planetary orbits. The dust-cloud mechanism that increases the likelihood of planets has also brought most stars into existence as double or multiple systems. It is obvious that the presence of two or more stars in a system will profoundly perturb the orbits of planets in the system. A few double-

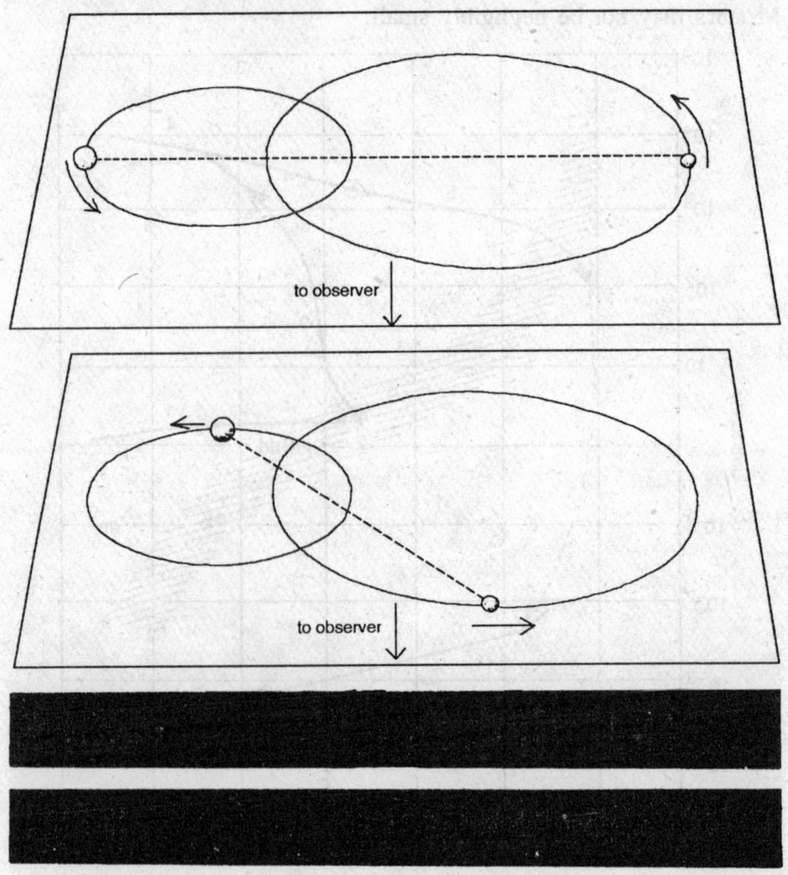

Fig. 3.8. Spectroscopic detection of close binaries is depicted in these diagram and spectra.

star systems have members that are so far apart that if they are sufficiently near to us, the stars can be distinguished by a telescope or even by the naked eye. Such systems are called visual binaries. Much more common are the "spectroscopic binaries." In these systems the stars are so close together that even if they are relatively near to us, they cannot be separated by the largest telescope. We can tell that they are double stars only by regular changes in their spectra. These changes also enable us to calculate their orbits.

The orbits of planets in such systems are so complicated that astronomers have been able to work them out for only a few idealized cases. A life-bearing planet would have to travel on an obrbit close to one of the stars in a widely separated system, or on an orbit at a large distance from the stars in a close system. In the case of a hypothetical binary composed of two stars with the same luminosity as our sun and revolving around a common center in a nearly circular orbit, a planet would find the thermally habitable zone dynamically stable only if the stars were more than 10 astronomical units or less than .05 astronomical unit apart (an astronomical unit is the distance between the earth and the sun). With a separation between the stars of .5 to 2 astronomical units there is no overlapping of the habitable zone and the dynamically stable zone. Therefore no inhabitable planets can exist in such systems. Taking everything into consideration, only 1 to 2 per cent of all double and multiple stars may possess inhabitable planets, and perhaps 3 to 5 per cent of all the stars in our galaxy have such planets.

For the present there is no hope of detecting on a photographic plate the existence of a planet of another star. Such planets as may attend even the nearest star are completely lost to view in the brilliance of the star's light. Some-day there may be a telescope on a platform in space, free of the interfering effects of the earth's atmosphere. As has been suggested by Nancy G. Roman of the National Aeronautics and Space Administration, the instrument would produce a sharp star image that could be blocked out so that a planet near the star could be detected by means of a long photographic exposure. Even the subtle methods used to detect spectroscopic binaries are not refined enough to find a planet, because the effect of a planet upon the motion of its parent star is so slight.

There is nonetheless other well-established evidence to support the contention that planets are common. In the first place, no sharp distinction can be drawn between binary or multiple stars and stars with planetary systems. According to Gerard P. Kuiper of the Yerkes Observatory, the mean distance of separation between the components of all binaries so far investigated is about 20 astronomical units. This is of the same order of magnitude as the distance between the sun and its major planets (Jupiter, Saturn, Uranus and Neptune). Kaj Aa. Strand at the U.S. Naval Observatory in Washington has studied small perturbations in the orbital motion of a star called 61 Cygni, which has a companion that is too faint to be directly observed. He has found that the mass of this unseen object is about a hundredth of that

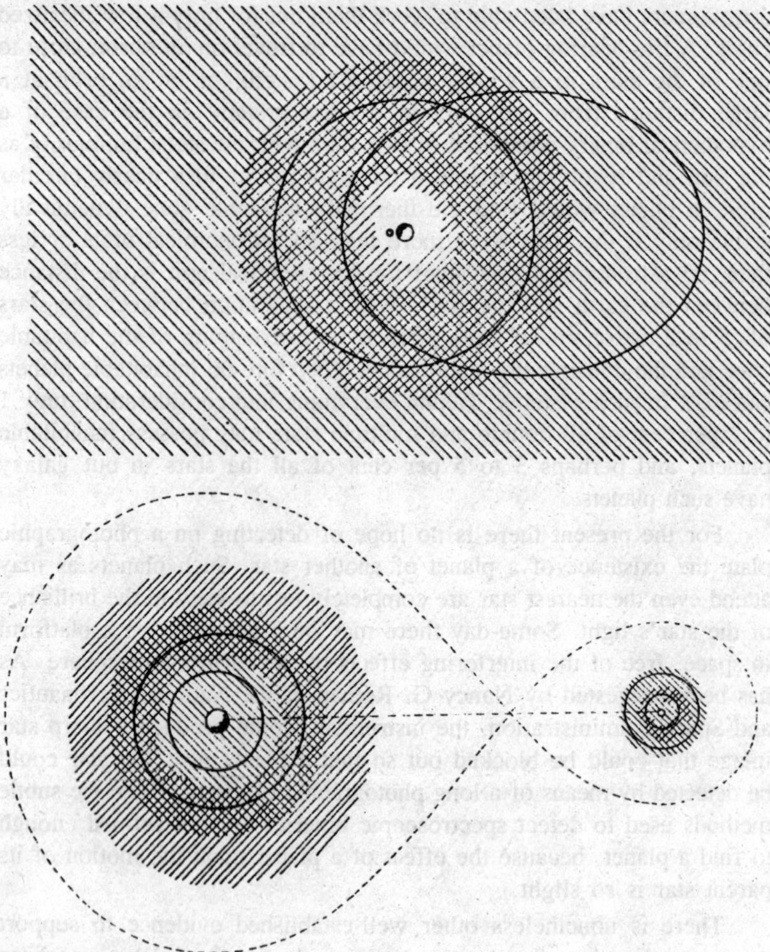

Fig. 3.9. Planetary orbits around binary stars are depicted here in an idealized manner.

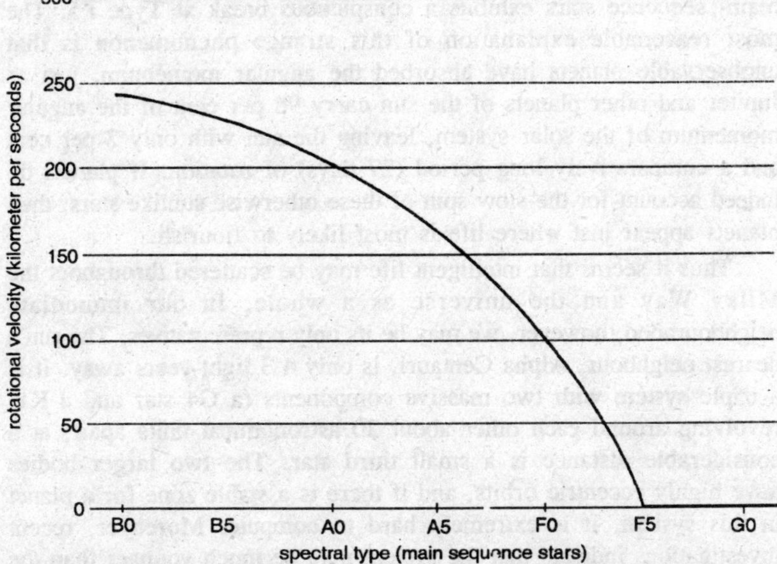

Fig. 3.10. Rotational velocity of stars on the main sequence is diagrammed here according to spectral types.

of the sun. This mass lies between that of stars and of Jupiter. It is therefore reasonable to believe that the masses of small stars in binary systems grade continuously down to the masses of planets. Since binaries are so common in our galaxy, it would seem that many stars that now appear to be alone actually possess planets.

Harold C. Urey of the Scripps Institution of Oceanography has found additional evidence for planets in a certain kind of meteorite. According to Urey, diamonds embedded in these objects show that they must at one time have been under high pressure in a body the size of the moon. Such moonlike objects, known as "pestellar nuclei," would enhance the formation of both stars and planets from a dust cloud. Any irregularity in the motions of a dust cloud should be expected to produce more than one such nucleus, and the formation of planets or multiple stars would follow as a normal consequence.

Finally, measurements of the angular momenta of many stars give every indication that planets exist outside the solar system. Otto Struve of the National Radio Astronomy Observatory has pointed out that main-sequence stars more massive than Type F5 usually rotate rapidly, but starting with this type the rotation of stars slows down abruptly. In other words, the average angular momentum per units mass of the main- sequence stars exhibits a conspicuous break at Type F5. The most reasonable explanation of this strange phenomenon is that unobservable planets have absorbed the angular momentum, just as Jupiter and other planets of the sun carry 98 per cent of the angular momentum of the solar system, leaving the sun with only 2 per cent and a comparatively long period (27 days) of rotation. If planets do indeed account for the slow spin of these otherwise sunlike stars, then planets appear just where life is most likely to flourish.

Thus it seems that intelligent life may be scattered throughout the Milky Way and the universe as a whole. In our immediate neighbourhood, however, we may be its only representatives. The sun's nearest neighbour, Alpha Centauri, is only 4.3 light-years away. It is a triple system with two massive components (a G4 star and a K1) revolving around each other about 20 astronomical units apart; at a considerable distance is a small third star. The two larger bodies have highly eccentric orbits, and if there is a stable zone for a planet in this system, it is extremely hard to compute. Moreover, recent investigations indicate that the system may be much younger than the sun, so that higher forms of life might not have had time to evolve even if a habitable planet does exist in it.

Forty other stars are located within five parsecs (16.7 light-years) of the sun. Only two-Epsilon Eridani (a K2 type) and Tau Ceti (G4) —seem to fulfil the conditions for the existence of advanced forms of life, and Epsilon Eridani may not be exactly on the main sequence. Tau Ceti is 10.8 light-years distant, has an apparent visual magnitude of 3.6, is located on the celestial sphere about 16 degrees south of its equator and appears above the horizon in the northern sky only in the winter.

Since intelligent life is probably not a rare phenomenon, and since at least one star in our vicinity meets the specifications of a life-fostering star, it may seem odd that we have had no visitors from other worlds. The idea would have drawn ridicule 20 years ago, but today it deserves consideration. There are, however, several reasons for believing that we have had no visitors from outer space. For one thing, the 10.8 light-years that separate us from Tau Ceti—astronomically a short distance—is an extremely long distance in terms of human experience. Even traveling at the speed of the artificial satellites that man has launched, space voyagers would need hundreds of thousands of years to traverse it. It is possible that organisms from Tau Ceti might have a far longer life-span than man, but this supposition invokes radical assumptions that cannot be supported by present knowledge. Furthermore, if a meeting is to occur, our high technological civilization would have to be contemporary with that created by intelligent organisms on other planets. The cultural evolution that brought mankind to its present technical competence began only a few centuries ago. And even if human civilization endures for hundreds of thousands of years, it would be a brief episode in the time-scale of biological evolution. Therefore the chance that advanced civilizations might be flourishing at the present time on the planets of one or two nearby stars is excessively small.

Some workers have concluded, however, that the chance is good enough and certainly intriguing enough—to institute radio surveillance of signals originating outside the solar system. Guiseppe Cocconi and Philip Morrison of Cornell University have pointed out that the most favourable wavelength would be one close to but outside of the 21-centimeter line emitted by hydrogen in space, because in this region of the spectrum the galactic noise and the noise produced in the earth's atmosphere are at a minimum Moreover, this important wavelength would be of as much interest to astronomers on another planet as to those on earth. Cocconi and Morrison have urged that radiation picked up by the 600-foot radio telescope, now being built by the Navy in West Virginia, be analyzed for the presence of signals. According to them, that telescope will be capable of detecting signals generated 10 light-years away by a technology no more advanced than our own. Meanwhile Frank D. Drake of the National Radio Astronomy Observatory, who makes a more generous estimate of the distance at which signals sent by intelligent organisms could be detected, is in charge of an actual project that is employing a smaller telescope to detect any radio signals transmitted by living beings in other "solar systems."

What kind of signals may we expect to receive or should we send out? Probably the most abstract and the most universal conception that any intelligent organisms anywhere would have devised is the sequence of cardinal numbers: 1, 2, 3, 4 and so on. The most likely signal would be a series of pulses indicating this sequence repeated at regular intervals. Such a signal may upon first consideration appear to be too simple for the sophisticated task of communicating with other beings far away among the stars. It would sound like baby talk. But after-all interstellar communication is surely still in the baby-talk stage.

4

World Beyond Earth

The Earth is but a small part of the world, and the rest of the world will be important to our future. In terms of energy, materials, and room for growth, space is almost everything. In the past, successes in space have regularly fulfilled engineering projections. In the future, an open space frontier will widen the human world. Advances in artificial intelligence and nanotechnology will play a crucial role. People took ages to recognize space as a frontier. Our ancestors once saw the night sky as a black dome with tiny sparks, a light show of the gods. They couldn't imagine space travel, because they didn't even know that outer space existed.

We now know that space exists, but few people yet understand its value. This is hardly surprising. Our minds and cultures have evolved on this planet, and we have just begun to digest the idea of a frontier beyond the sky.

Project Apollo

Only in this century did such visionary designers as Hermann Oberth and Robert Goddard show that rockets could reach space. They had confidence in this because they knew enough about fuel, engines, tankage, and structures to calculate what multistage rockets could do. Yet, in 1921 a New York Times editorialist chided Goddard for the notion that rockets could fly through space without air to push against, and as late as 1956 the Astronomer Royal of Britain snorted that "*Space travel is utter bilge.*" This only showed that editorialists and astronomers were the wrong experts to ask about space hardware. In 1957, Sputnik orbited Earth, followed in 1961 by Yuri Gagarin. In 1969, the world saw footprints on the Moon.

We paid a price for ignorance, though. Because the pioneers of space technology had lacked any way to establish their case in public, they were forced to argue basic points again and again ('Yes, rockets will work in vacuum... Yes, they really will reach orbit....). Busy defending the basics of spaceflight, they had little time to discuss its consequences. Thus, when Sputnik startled the world and embarrassed the United States, people were unprepared: there had been no widespread debate to shape a strategy for space.

Some of the pioneers had seen what to do: build a *space station* and a reusable spaceship, then reach out to the Moon or asteroids for resources. But the noise of flustered politicians promptly drowned out their suggestions, and U.S. politicians clamored for a big, easy-to-understand goal. Thus was born *Project Apollo*, the race to land a U.S. citizen on the nearest place to plant a flag. *Project Apollo* by passed building a space station and space shuttle, instead building giant missiles able to reach the Moon in one great leap. The project was glorious, it gave scientists some information, and it brought great returns through advances in technology—but at the core, it was a hollow stunt. Taxpayers saw this, congressmen saw this, and the space program shriveled.

During Apollo, old dreams held sway in the public mind, and they were simple, romantic dreams of settling other planets. Then robot instruments dissolved the dream of a jungle-clad Venus in the reality of a planet-wide oven of high-pressure poison. They erased the lines Earthbound astronomers had drawn on Mars, and with them went both canals and Martians. In their place was a Mars of craters and canyons and dry blowing dust. Sunward of Venus lay the baked rock of Mercury; starward of Mars lay rubble and ice. The planets ranged from dead to murderous, and the dream of new Earths receded to distant stars. Space seemed a dead end.

New Space Program

A new space program has risen from the ruin of the old. A new generation of space advocates, engineers, and entrepreneurs now aims to make space the frontier it should have been from the beginning—a place for development and use, not for empty political gestures. They have confidence in success because space development requires no breakthrough in science or technology. Indeed, the human race could conquer space by applying the technologies of twenty years ago—and by avoiding stunt flights, we could probably do it at a profit. Space activities need not be expensive.

Consider the high cost of reaching orbit today—thousands of dollars per kilogram. Where does it come from? To a spectator at a *shuttle launch*, shaken by the roar and awed by the flames, the answer seems obvious: the fuel must cost a mint. Even airlines pay roughly half their direct operating costs for fuel. A rocket resembles an airliner—it is made of aluminum and stuffed with engines, controls, and electronics—but fuel makes up almost all its mass as it sits on the launch pad. Thus, one might expect fuel to account for well over half the operating cost of a rocket. But this expectation is false. In the Moon shots, the cost of the fuel needed to reach orbit amounted to less than a million dollars—a few dollars per kilogram delivered to orbit, a fraction of a percent of the total cost. Even today, fuel remains a negligible part of the cost of spaceflight.

Why is spaceflight so much costlier than air flight? In part, because spacecraft aren't made in quantity; this forces manufacturers to re cover their design costs from sales of only a few units, and to make those few units by hand at great cost. Further, most spacecraft are thrown away after one use, and even shuttles are flown just a few times a year—their cost cannot be spread over several flights a day for years, as the cost of airliners can. Finally, spaceport costs are now spread over only a few flights per month, when large airports can spread their cost over many thousands. All this conspires to make each flight into space dauntingly expensive.

But studies by Boeing Aerospace Company—the people who brought inexpensive jet transportation to much of the world—show that a fleet of fully reusable shuttles, flown and maintained like airliners, would drop the cost of reaching orbit by a factor of fifty or more. The key is not new technology, but economies of scale and changes in management style.

Communications Satellites

Space offers vast industrial opportunities. The advantages of perching observation and communications satellites on orbit are well known. Future *communications satellites* will be powerful enough to communicate with hand-held stations on the ground, bringing the ultimate in mobile telephone service. Companies are already moving to take advantage of zero gravity to perform delicate separation processes, to make improved pharmaceuticals; other companies plan to grow better electronic crystals. In the years before assemblers take over materials production, engineers will use the space environment to extend the abilities of bulk technology. Space industry will pro vide a growing market for launch services, dropping launch costs. Falling launch costs, in turn, will stimulate the growth of space industry. Rocket transportation to Earth orbit will eventually become economical.

Space planners and entrepreneurs are already looking beyond Earth orbit to the resources of the *solar system*. In deep space, how ever, rockets swiftly become too expensive for hauling freight—they gobble fuel that itself had to be hauled into space by rockets. Fuel-burning rockets are as old as Chinese fireworks, far older than "*The Star-Spangled Banner*." They evolved for natural reasons: compact, powerful, and useful to the military, they can punch through air and fight strong gravity. *Space engineers* know of alternatives, however.

Vehicles need no great blasts of power to move through the frictionless vacuum of space. Small forces can slowly and steadily push a vehicle to enormous speeds. Because energy has mass, sunlight bouncing off a thin mirror—a solar sail—provides such a force. The pull of solar gravity provides another. Together, light pressure and gravity can carry a spacecraft anywhere in the solar system and back again. Only the heat near the Sun and the drag of planetary atmospheres will limit travel, forcing sails to steer clear of them.

NASA has studied solar sails designed to be carried to space in rockets, but these must be fairly heavy and sturdy to survive the stress of launch and unfolding. Eventually, engineers will make sails in space, using a low-mass tension structure to support mirrors of thin metal film. The result will be the "*light sail*," a higher-performance class of solar sail. After a year's acceleration, a light sail can reach a speed of one hundred kilometers per second, leaving today's swiftest rockets in the dust.

If you imagine a network of graphite-fiber strands, a spinning spider web kilometers wide with gaps the size of football fields between the strands, you will be well on your way to imagining the structure of a light sail. If you picture the gaps bridged by reflecting panels built of aluminum foil thinner than a soap bubble, you will have a fair idea of how it looks: many reflective panels tied close together to form a vast, rippled mosaic of mirror. Now picture a load of cargo hanging from the web like a parachutist from a parachute, while centrifugal force holds the web-slung mirror taut and flat in the void, and you almost have it.

To build light sails with bulk technology, we must learn to make them in space; their vast reflectors will be too delicate to survive launch and unfolding. We will need to construct scaffolding structures, manufacture thin-film reflectors, and use remotely controlled robot arms in space. But space planners already aim to master construction, manufacturing, and robotics for other space applications. If we build light sails early in the course of space development, the effort will exercise these skills without requiring the launch of much material. Though vast, the scaffolding (together with materials for many sails) will be light enough for one or two shuttle flights to lift to orbit.

Sailing

A *sail* production facility will produce sails cheaply. The sails, once built, will be cheap to use: they will have few critical moving parts, little mass, and zero fuel consumption. They will be utterly different from rockets in form, function, and cost of operation. In fact, calculations suggest that the costs will differ by a factor of roughly a thou sand, in favour of lightsails.

Today most people view the rest of the solar system as vast and inaccessible. It is vast; like the Earth, it will take months to circumnavigate by sail. Its apparent inaccessibility, however, has less to do with distance than with the cost of transportation via rocket.

Light sails can smash the cost barrier, opening the door to the solar system. Light sails will make other planets easier to reach, but this will not make planets much more useful: they will remain deadly deserts. The gravity of planets will prevent light sails from shuttling to their surfaces, and will also handicap industry on a planet's surface. Spinning space stations can simulate gravity if it is needed, but planet-bound stations cannot escape it. Worse yet, planetary atmospheres block solar energy, spread dust, corrode metals, warm refrigerators, cool ovens, and blow things down. Even the airless Moon rotates, blocking sunlight half the time, and has gravity enough to ground light sails beyond hope of escape. Light sails are fast and tireless, but not strong.

The great and enduring value of space lies in its resources of matter, energy, and room. The planets occupy room and block energy. The material resources they offer are inconveniently placed. The asteroids, in contrast, are flying mountains of resources that trace orbits crisscrossing the entire solar system. Some cross the orbit of Earth; some have even struck Earth, blasting craters. Mining the asteroids seems practical. We may need roaring rockets to carry things up into space, but meteorites prove that ordinary rocks can fall down from space—and like the space shuttle, objects falling from space need not burn up on the way down. Delivering packages of material from an asteroid to a landing target in a salt flat will cost little.

Sources and Resources

Even small asteroids are big in human terms: they hold billions of tons of resources. Some asteroids contain water and a substance resembling oil shale. Some contain fairly ordinary rock. Some contain a metal that holds elements scarce in Earth's crust, elements that sank beyond reach ages ago in the formation of Earth's metal core: this meteoritic steel is a strong, tough alloy of iron, nickel, and cobalt, bearing valuable amounts of platinum-group metals and gold. A kilometer-wide chunk of this material (and there are many) contains precious metals worth several trillion dollars, mixed with enough nickel and cobalt to supply Earth's industry for many years.

The Sun floods space with easily collected energy. A square-kilometer framework holding metal-film reflectors will gather over a billion watts of sunlight, free of interference from cloud or night. In the weather less calm of space, the flimsiest collector will be as permanent as a hydroelectric dam. Since the Sun puts out as much energy in a microsecond as the human race now uses in a year, energy need not be scarce for some time to come.

Finally, space itself offers room to live. People once saw life in space in terms of planets. They imagined domed cities built on planets, dead planets slowly converted into Earth-like planets, and Earth-like planets reached after years in a flight to the stars. But planets are package deals, generally offering the wrong gravity, atmosphere, length of day, and location.

Free space offers a better building site for settlements. Professor Gerard O'Neill of Princeton University brought this idea to public attention to, helping to revive interest in space after the post-Apollo crash. He showed that ordinary construction materials—and steel and glass—could be used to build habitable cylinders in space, kilometers in length and circumference. In his design, dirt underfoot shields inhabitant from the natural radiation of space, just as Earth's inhabit ants are shielded by the air overhead. Rotation produces an acceleration equaling Earth's gravity, and broad mirrors and window panels flood the interior with sunlight. Add soil, streams, vegetation, and imagination, and the lands inside could rival the best valleys on Earth its places to live. With just the resources of the asteroids, we will be able to build the practical equivalent of a thousand new Earths.

By adapting present technology, we could open the space frontier. The prospect is heartening. It shows us an obvious way to bypass terrestrial limits to growth, lessening one of the fears that has clouded our view of the future. The promise of the space frontier can thus mobilize human hope—a resource we will need in abundance, if we are to deal with other problems.

Space and Advanced Technology

By adapting present technology, we could indeed open the space frontier—but we won't. Along the path foreseen by the current space movement, human civilization would take decades to become firmly established in space. Before then, breakthroughs in technology will open new paths.

Nowadays, teams of engineers typically take five to ten years to develop a new space system, spending tens to thousands of millions of dollars along the way. These engineering delays and costs make progress painfully slow. In coming years, though, computer-aid design systems will evolve toward automated engineering systems. As they do, engineering delays and costs will shrink and then plummet; computer-controlled manufacturing systems will drop overall costs still further. A day will come when automated design and manufacturing will have made space systems development more than tenfold faster and cheaper. Our progress in space will soar.

At that time, will space settlers look back on our present space program as the key to space development? Perhaps not. They will have seen more technical progress made in a few years than space engineers previously managed in a few decades. They may well conclude that AI and robotics did more for space development than did a whole army of NASA engineers.

The assembler breakthrough and automated engineering will combine to bring advances that will make our present space efforts seem quaint. As described already that replicating assemblers will be able to build a light, strong rocket engine using little human labour. Using similar methods, we will build entire spacecraft of low cost and extraordinary performance. Weight for weight, their diamond-based structural materials will have roughly fifty times the strength (and fourteen times the stiffness) of the aluminum used in the present shuttle; vehicles built with these materials can be made over 90 percent lighter than similar vehicles today. Once in space, vehicles will spread solar collectors to gather abundant energy. Using this energy to power assemblers and disassemblers, they will rebuild themselves in Bight to suit changing conditions or the whim of their passengers. Today, space travel is a challenge. Tomorrow, it will be easy and convenient.

Since *nanotechnology* lends itself to making small things, consider the smallest person-carrying spacecraft: the spacesuit. Forced to use weak, heavy, passive materials, engineers now make bulky, clumsy spacesuits. A look at an advanced spacesuit will illustrate some of the capabilities of nanotechnology.

Imagine that you are aboard a space station, spun to simulate Earth's normal gravity. After instruction, you have been given a suit to try out: there it hangs on the wall, a gray, rubbery-looking thing with a transparent helmet. You take it down, heft its substantial weight, strip, and step in through the open seam on the front.

The suit feels softer than the softest rubber, but has a slick inner surface. It slips on easily and the seam seals at a touch. It provides a skintight covering like a thin leather glove around your fingers, thickening as it runs up your arm to become as thick as your hand in the region around your torso. Behind your shoulders, scarcely noticeable, is a small backpack. Around your head, almost invisible, is the helmet. Below your neck the suit's inner surface hugs your skin with a light, uniform touch that soon becomes almost imperceptible.

You stand up and walk around, experimenting. You bounce on your toes and feel no extra weight from the suit. You bend and stretch and feel no restraint, no wrinkling, no pressure points. When you rub your fingers together they feel sensitive, as if bare—but somehow slightly *thicker*, As you breathe, the air tastes clean and fresh. In fact, you feel that you could forget that you are wearing a suit at all. What is more, you feel just as comfortable when you step out into the vacuum of space.

The suit manages to do all this and more by means of complex activity within a structure having a texture almost as intricate as that of living tissue. A glove finger a millimeter thick has room for a thousand micron-thick layers of active nanomachinery and nanoelectronics. A fingertip-sized patch has room for a billion mechanical nano computers, with 99.9 percent of the volume left over for other components.

In particular, this leaves room for an active structure. The middle layer of the suit material holds a three-dimensional weave of diamond-based fibers acting much like artificial muscle, but able to push as well as pull. These fibers take up much of the volume and make the suit material as strong as steel. Powered by microscopic electric motors and controlled by nanocomputers, they give the suit material its supple strength, making it stretch, contract, and bend as needed. When the suit felt soft earlier, this was because it had been programmed to act soft. The suit has no difficulty holding its shape in a vacuum; it has strength enough to avoid blowing up like a balloon. Likewise, it has no difficulty supporting its own weight and moving to match your motions, quickly, smoothly, and without resistance. This is one reason why it almost seems not to be there at all.

Your fingers feel almost bare because you feel the texture of what you touch. This happens because pressure sensors cover the suit's surface and active structure covers its lining: the glove feels the shape of whatever you touch—and the detailed pattern of pressure it exerts—and transmits the same texture pattern to your skin. It also reverses the process, transmitting to the outside the detailed pattern of forces exerted by your skin on the inside of the glove. Thus the glove pretends that it isn't there, and your skin feels almost bare.

The suit has the strength of steel and the flexibility of your own body. If you reset the suit's controls, the suit continues to match your motions, but with a difference. Instead of simply transmitting the forces you exert, it amplifies them by a factor of ten. Likewise, when something brushes against you, the suit now transmits only a tenth of the force to the inside. You are now ready for a wrestling match with a gorilla.

The fresh air you breathe may not seem surprising; the backpack includes a supply of air and other consumables. Yet after a few days outside in the sunlight, your air will not run out: like a plant, the suit absorbs sunlight and the carbon dioxide you exhale, producing fresh oxygen. Also like a plant (or a whole ecosystem), it breaks down other wastes into simple molecules and reassembles them into the molecular patterns of fresh, wholesome food. In fact, the suit will keep you comfortable, breathing, and well fed almost anywhere in the inner solar system.

What is more, the suit is durable. It can tolerate the failure of numerous nanomachines because it has so many others to take over the load. The space between the active fibers leaves room enough for assemblers and disassemblers to move about and repair damaged devices. The suit repairs itself as fast as it wears out.

Within the bounds of the possible, the suit could have many other features. A speck of material smaller than a pinhead could hold the text of every book ever published, for display on a fold-out screen. Another speck could be a "seed" containing the blueprints for a range of devices greater than the total the human race has yet built, along with replicating assemblers able to make any or all of them.

What is more, fast technical artificial intelligence systems could design the suit in a morning and have it built by afternoon. All that we accomplish in space with modern bulk technology will be swiftly and dramatically surpassed shortly after molecular technology and automated engineering arrive. In particular, we will build replicating assemblers that work in space. These replicators will use solar energy as plants do, and with it they will convert asteroidal rubble into copies of themselves and products for human use, With them, we will grasp the resources of the solar system.

By now, most readers will have noted that this, like certain earlier discussions, sounds like science fiction. Some may be pleased, some dismayed that future possibilities do in fact have this quality. Some, though, may feel that "*sounding like science fiction*" is somehow grounds for dismissal. This feeling is common and deserves scrutiny.

Technology and science fiction have long shared a curious relation ship. In imagining future technologies, SF writers have been guided partly by science, partly by human longings, and partly by the market demand for bizarre stories. Some of their imaginings later become real, because ideas that seem plausible and interesting in fiction sometimes prove possible and attractive in actuality. What is more, when scientists and engineers foresee a dramatic possibility, such as rocket spaceflight, SF writers commonly grab the idea and popularize it.

Later, when engineering advances bring these possibilities closer to realization, other writers examine the facts and describe the prospects. These descriptions, unless they are quite abstract, then sound like science fiction. Future possibilities will often resemble today's fiction, just as robots, spaceships, and computers resemble yesterday' fiction. How could it be otherwise? Dramatic new technologies sound like science fiction because science fiction authors, despite their frequent fantasies, aren't blind and have a professional interest in the area.

Science fiction authors often *fictionalize* (that is, counterfeit) the scientific content of their stories to "explain" dramatic technologies. Some fuzzy thinkers then take all descriptions of dramatic technical advances, lump them together with this bogus science, and ignore the lot. This is unfortunate. When engineers project future abilities, they test their ideas, evolving them to fit our best understanding of the laws of nature. The resulting concepts must be distinguished from ideas evolved to fit the demands of paperback fiction. Our lives will depend on it. Much will remain impossible, even with molecular technology. No spacesuit, however marvelous, will be able to rocket back and forth indefinitely at tremendous speeds, or survive great explosions, or walk through walls, or even stay cool indefinitely in a hot isolated room. We have far to go before reaching the limits of the possible, yet limits exist. But this is a topic taken up later.

ABUNDANCE

Space resources join with assemblers and automated engineering systems to round out the case for a future of great material abundance. What this means can best be seen by examining costs.

Costs reflect the limits of our resources and abilities; high costs indicate scarce resources and difficult goals. The prophets of scarcity have in effect predicted steeply rising resource costs, and with them a certain kind of future. *Resource costs*, however, always depend on technology. Unfortunately, engineers attempting to predict the cost of future technologies have generally encountered a tangle of detail and uncertainty that proves impossible to untie. This problem has obscured our understanding of the future.

The prospect of replicating assemblers, automated engineering, and space resources cuts this Gordian knot of cost prediction. Today the cost of products includes the costs of labour, capital, raw materials, energy, land, waste disposal, organization, distribution, taxation, and design. To see how total costs will change, consider these elements one by one.

Labour

Replicating assemblers will require no labour to build, once the first exists. What use are human hands in running an assembler? Further, with robotic devices of various sizes to assemble parts into larger systems, the entire manufacturing process from assembling molecules to assembling skyscrapers could be free of labour costs.

Capital

Assembler-based systems, if properly programmed, will themselves be productive capital. Together with larger robotic machines, they will be able to build virtually anything, including copies of themselves. Since this self-replicating capital will be able to double many times per day, only demand and available resources will limit its quantity. Capital as such need cost virtually nothing.

Raw Materials

Since *molecular machines* will arrange atoms to best advantage, a little material can go a long way. Common elements like hydrogen, carbon, nitrogen, oxygen, aluminum, and silicon seem best for constructing the bulk of most structures, vehicles, computers, clothes and so forth: they are light and form strong bonds. Because dirt and air contain these elements in abundance, raw materials can be dirt cheap.

Energy

Assemblers will be able to run off chemical or electrical energy. *Assembler-built systems* will convert solar to chemical energy, like plants, or solar to electrical energy, like solar cells. Existing solar cells are already more efficient than plants. With replicating assemblers to build solar collectors, fuel and electric power will cost little.

Land

Assembler-based production systems will occupy little room. Most could sit in a closet (or a thimble, or a pinhole); larger systems could be placed underground or in space if someone wants something that requires an unsightly amount of room. Assembler- based production systems will make both digging machines and spacecraft cheap.

Waste Disposal

Assembler systems will be able to keep control of the atoms they use, making production as clean as a growing apple tree, or cleaner. If the orchard remains too dirty or ugly, we will be able to move it off Earth entirely.

Organization

Today, factory production requires *organization* to coordinate hordes of workers and managers. *Assembler-based* production machines will contain no people, and will simply sit around and produce things made to order. Their initial programming will provide all the organization and information needed to make a wide range of products.

Distribution

With automatic vehicles running in tunnels made by cheap digging machines, distribution need neither consume labour nor blight the landscape. With assemblers in the home and community, there will be less need for distribution in the first place.

Taxation

Most taxes take a fixed percentage of a price, and thus add a fixed percentage to the cost. If the cost is negligible, the tax will be negligible. Further, governments with their own replicators and raw materials will have less reason to tax people.

Design

The above points add up to a case for low costs of production. Technical AI systems, by avoiding the labour cost of engineering, will virtually eliminate the costs of design. These AI systems will themselves be inexpensive to produce and operate, being constructed by assemblers and having no inclination to do anything but design things.

In short, at the end of a long line of profitable developments in computer and molecular technologies, the cost of designing and producing things will drop dramatically. I above referred to "dirt cheap" raw materials, and indeed, assemblers will be able to make almost anything from dirt and sunlight. Space resources, however, will change "*dirt cheap*" to "*cheap-dirt cheap*": topsoil has value in Earth's ecosystem, but rubble from asteroids will come from a dead and dreary desert. By the same token, assemblers in space will run off cheap sunlight.

Space resources are vast. One asteroid could bury Earth's continents a kilometer deep in raw materials. Space swallows the 99.999999955 percent of the Sun's light that misses Earth, and most is lost to the interstellar void.

Space holds matter, energy, and room enough for projects of vast size, including vast space settlements. Replicator-based systems will be able to construct worlds of continental scale, resembling Dr. O'Neill's cylinders but made of strong, carbon-based materials. With these materials and water from the ice moons of the outer solar system, we will be able to create not only lands in space, but whole seas, wider and deeper than the Mediterranean. Constructed with energy and materials from space, these broad new lands and seas will cost Earth and its people almost nothing in terms of resources. The chief requirement will be programming the first replicator, but AI systems will help with that. The greatest problem will be deciding what we want.

As Konstantin Tsiolkovsky wrote near the turn of the century, "Man will not always stay on Earth; the pursuit of light and space will lead him to penetrate the bounds of the atmosphere, timidly at first, but in the end to conquer the whole of solar space." To dead space we will bring life.

And replicators will give us the resources to reach for the stars. A light sail driven starward only by sunlight would soon find itself coasting in the dark—faster than any modern rocket, yet so slowly that it would take millennia to cross the interstellar gulf. We can build a tremendous bank of lasers orbiting the Sun, however, and with it drive a beam far beyond our solar system, pushing a sail toward the speed of light. The crossing then will take only years.

Stopping presents a problem. Freeman Dyson of Princeton suggests braking with magnetic fields in the thin ionized gas between the stars. Robert Forward of Hughes Research Laboratories suggests bouncing laser light off the sail, directing light back along the sail's path to decelerate a smaller sail trailing behind. One way or another (and there are many others), the stars themselves lie within our reach.

For a long time to come, however, the solar system can provide room enough. The space near Earth holds room for lands with a million times Earth's area. Nothing need stop emigration, or return visits to the old country. We will have no trouble powering the transportation system—the sunlight falling on Earth supplies enough energy in ten minutes to put today's entire population in orbit. Space travel and space settlements will both become cheap. If we make wise use of molecular technology, our descendants will wonder what kept us bottled up on Earth for so long, and in such poverty.

Positive-sum Society

It might seem that the cost of everything—even land, if one does not crave thousands of kilometers of rock underfoot—Will drop to nothing. In a sense, this is almost right; in another sense, it is quite false. People will always value matter, energy, information, and genuine human service, therefore everything will still have its cost. And in the long run, we will face real limits to growth, so the cost of resources cannot be dismissed.

Nonetheless, if we survive, *replicators* and *space resources* will bring a long era in which genuine resource limits do not yet pinch us—an era when by our present standards even vast wealth will seem virtually free. This may seem too good to be true, but nature (as usual) has not set her limits based on human feelings. Our ancestors once thought that talking to someone across the sea (many months' voyage by sailing ship) would be too good to be true, but undersea cables and oversea satellites worked anyway.

But there is another, less pleasant answer for those who think assemblers are too good to be true: assemblers also threaten to bring hazards and weapons more dangerous than any yet seen. If *nanotechnology* could be avoided but not controlled, then sane people would shun it. The technology race, however, will bring forth assemblers from *biotechnology* as surely as it brought forth spacecraft from missiles. The military advantages alone will be enough to make advances almost inevitable. Assemblers are unavoidable, but perhaps controllable.

Our challenge is to avoid the dangers, but this will take cooperation, and we are more likely to cooperate if we understand how much we have to gain from it. The prospect of space and replicating assemblers may help us clear away some ancient and dangerous meters.

Human life 'was once like a zero-sum game. Humankind lived near its ecological limit and tribe fought tribe for living space. Where pastures, farmland, and hunting grounds were concerned, more for one group meant less for another. Because one's gain roughly equaled the other's loss, net benefits summed to zero. Still, people who cooperated on other matters prospered, and so our ancestors learned not just to grab, but to cooperate and build.

Where taxes, transfer payments, and court battles are concerned, more for one still means less for another. We add to total wealth slowly, but redistribute it swiftly. On any given day our resources seem fixed, and this gives rise to the illusion that life is a zero-sum game. This illusion suggests that broad cooperation is pointless, be cause our gain must result from some opponent's loss. The history of human advance proves that the world game can be positive-sum. Accelerating economic growth during recent centuries shows that the rich can get richer while the poor get richer. Despite population growth (and the idea of dividing a fixed pie) the average wealth per capita worldwide, including that of the *Third World*, has grown steadily larger. Economic fluctuations, local reversals, and the natural tendency of the media to focus on bad news—these combine to obscure the facts about economic growth, but public records show it clearly enough. Space resources and replicating assemblers will accelerate this historic trend beyond the dreams of economists, launching the human race into a new world.

5

PRINCIPLES OF CHANGE

Molecular assemblers will bring a revolution without parallel since the development of *ribosomes*, the primitive assemblers in the cell. The resulting nanotechnology can help life spread beyond Earth—a step without parallel since life spread beyond the seas. It can help mind emerge in machines—a step without parallel since mind emerged in primates. And it can let our minds renew and remake our bodies—a step without any parallel at all.

These revolutions will bring dangers and opportunities too vast for the human imagination to grasp. Yet the principles of change that have applied to molecules, cells, beasts, minds, and machines should endure even in an age of *biotechnology*, *nanomachines*, and *artificial minds*. The same principles that have applied at sea, on land, and in the air should endure as we spread Earth's life toward the stars. Understanding the enduring principles of change will help us under stand the potential for good and ill in the new technologies.

ORDER FROM CHAOS

Order can emerge from chaos without anyone's giving orders: orderly crystals condensed from formless interstellar gas long before Sun, Earth, or life appeared. Chaos also gives rise to a crystalline order under more familiar circumstances. Imagine a molecule—perhaps regular in form, or perhaps lopsided and knobby like a ginger root. Now imagine a vast number of such molecules moving randomly in a liquid, tumbling and jostling like drunkards in weight less ness in the dark. Imagine the liquid evaporating and cooling, forcing the molecules closer together and slowing them down. Will these randomly moving, oddly shaped molecules simply gather in disordered heaps? Generally not. They will usually settle into a crystalline pattern, each neatly nestled against its neighbours, forming rows and columns as perfect as a checkerboard, though often more complex.

This process involves neither magic nor some special property of molecules and quantum mechanical forces. It does not even require the special matching shapes that enable protein molecules to self-assemble into machines. Marbles of uniform size, if placed in a tray and shaken, also settle into a regular pattern.

Crystals grow by trial and the removal of error, by variation and selection. No tiny hands assemble them. A crystal can begin with a chance clumping of molecules: the molecules wander, bump, and clump at random, but clumps stick best when packed in the right crystalline pattern. Other molecules then strike this first, tiny crystal. Some bump in the wrong position or orientation; they stick poorly and shake loose again. Others happen to bump properly; they stick better and often stay. Layer builds on layer, extending the crystalline pattern. Though the molecules bump at random, they do not stick at random. Order grows from chaos through variation and selection.

Evolving Molecules

In *crystal growth*, each layer forms a template for the next. Uniform layers accumulate to form a solid block.

In cells, strands of DNA or RNA can serve as templates too, aided by enzymes that act as molecular copying machines. But the subunits of nucleic acid strands can be arranged in many different sequences, and a template strand can separate from its copy. Both strand and copy can then be copied again. Biochemist Sol Spiegelman has used a copying machine (a protein from a virus) in test tube experiments. In a simple, lifeless environment, it duplicates RNA molecules.

Picture a strand of RNA floating in a test tube, together with copying machines and RNA subunits. The strand tumbles and writhes until it bumps into a copying machine in the right position to stick. Subunits bump around until one of the right kind meets the copying machine in the right position to match the template strand. As matching subunits chance to fall into position, the machine seizes them and bonds them to the growing copy; though subunits bump randomly, the machine bonds selectively. Finally the machine, the template, and the copy separate.

In the terminology of Oxford zoologist Richard Dawkins, things that give rise to copies of themselves are called *replicators*. In this

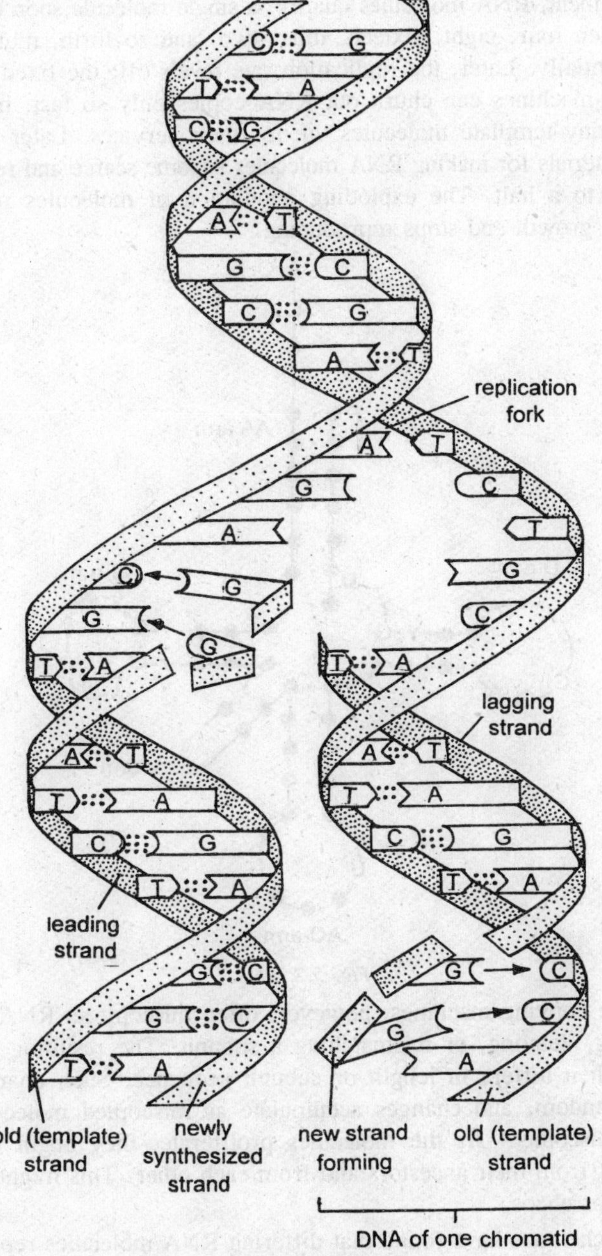

Fig. 5.1. Replication of DNA.

environment, RNA molecules qualify: a single molecule soon be comes two, then four, eight, sixteen, thirty-two, and so forth, multi plying exponentially. Later, the replication rate levels off: the fixed stock of protein machines can churn out RNA copies only so fast, no matter how many template molecules vie for their services. Later still, the raw materials for making RNA molecules become scarce and replication starves to a halt. The exploding population of molecules reaches a limit to growth and stops reproducing.

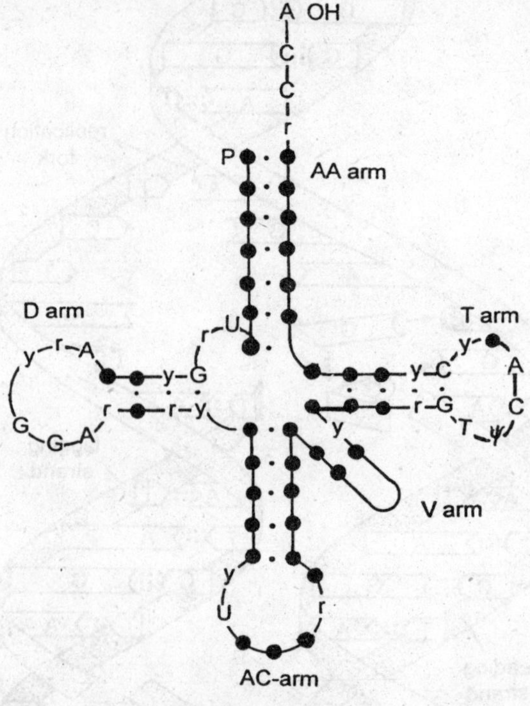

Fig. 5.2. RNA.

The copying machines, however, often miscopy an RNA strand, inserting, deleting, or mismatching a subunit. The resulting mutated strand then differs in length or subunit sequence. Such changes are fairly random, and changes accumulate as miscopied molecules are again miscopied. As the molecules proliferate, they begin to grow different from their ancestors and from each other. This might seem a recipe for chaos.

Biochemists have found that differing RNA molecules replicate at differing rates, depending on their lengths and subunit patterns.

Descendants of the swifter replicators naturally grow more common. Indeed, if one kind replicates just 10 percent more rapidly than its siblings, then after one hundred generations, each of the faster kind gives rise to 1,000 times as many descendants. Small differences in exponential growth pile up exponentially.

When a test tube runs out of subunits, an experimenter can sample its RNA and "*infect*" a fresh tube. The process begins again and the molecules that dominated the first round of competition begin with a head start. More small changes appear, building over time into large changes. Some molecules replicate faster, and their kind dominates the mix. When resources run out, the experimenter can sample the RNA and start again (and again, and again), holding conditions stable.

This experiment reveals a natural process: no matter what RNA sequences the experimenter starts with, the seeming chaos of random errors and biased copying brings forth one kind of RNA molecule (give or take some copying errors). Its typical version has a known, well-defined sequence of 220 subunits. It is the best RNA replicator in this environment, so it crowds out the others and stays.

Prolonged copying, miscopying, and competition always bring about the same result, no matter what the length or pattern of the RNA molecule that starts the process. Though no one could have predicted this winning pattern, anyone can see that change and competition will tend to bring forth a single winner. Little else could happen in so simple a system. If these replicators affected one an other strongly (perhaps by selectively attacking or helping one an other), then the result could resemble a more complex ecology. As it is, they just compete for a resource.

A variation on this example shows us something else: RNA molecules adapt differently to different environments. A *molecular machine* called a *ribonuclease* grabs RNA molecules having certain sequences of exposed subunits and cuts them in two. But RNA molecules, like proteins, fold in patterns that depend on their sequences, and by folding the right way they can protect their vulnerable spots. Experimenters find that RNA molecules evolve to sacrifice swift replication for better protection when ribonuclease is around. Again, a best competitor emerges.

Notice that biological terms have crept into this description: since the molecules replicate, the word "*generation*" seems right; molecules "*descended*" from a common "*ancestor*" are "*relatives*," and the words "*growth*," "*reproduction*," "*mutation*," and "*competition*" also seem right. Why is this? Because these molecules copy themselves with small variations, as do the genes of living organisms. When varying replicators have varying successes, the more successful tend to accumulate. This process, wherever it occurs, is "*evolution.*"

In this test tube example we can see evolution stripped to its bare essentials, free of the emotional controversy surrounding the evolution of life. The RNA replicators and protein copying machines are well-defined collections of atoms obeying well-understood principles and evolving in repeatable laboratory conditions. Biochemists can make RNA and protein from off-the-shelf chemicals, without help from life.

Biochemists borrow these *copying machines* from a kind of *virus* that infects bacteria and uses RNA as its genetic material. These viruses survive by entering a bacterium, getting themselves copied

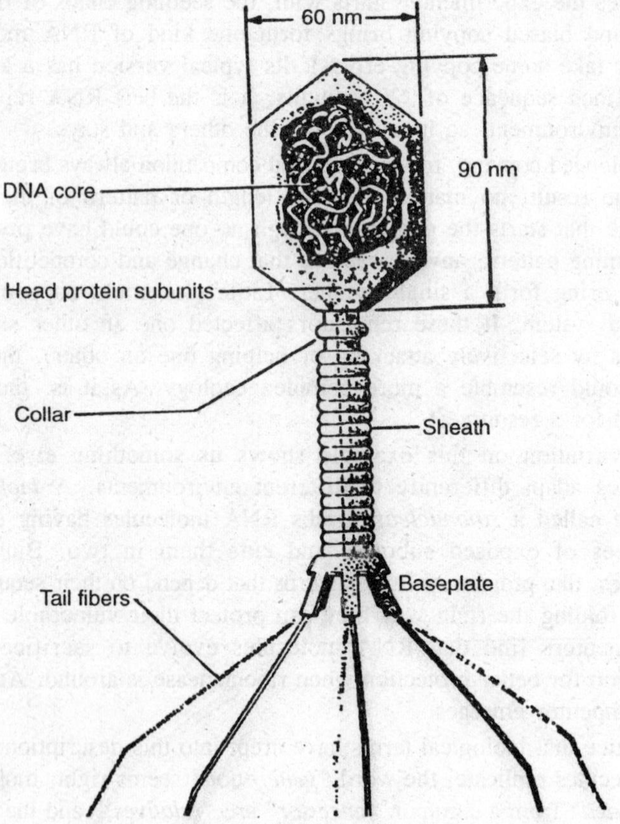

Fig. 5.3. Bacteriophage.

using its resources, and then escaping to infect new bacteria. Miscopying of viral RNA produces mutant viruses, and viruses that replicate more successfully grow more common; this is evolution by natural selection, apparently called "*natural*" because it involves nonhuman parts of nature. But unlike the test tube RNA, viral RNA must do more than just replicate itself as a bare molecule. Successful viral RNA must also direct bacterial ribosomes to build protein devices that let it first escape from the old bacterium, then survive outside, and finally enter a new one. This additional information makes viral RNA molecules about 4,500 subunits long.

To replicate successfully, the DNA of large organisms must do even more, directing the construction of tens of thousands of different protein machines and the development of complex tissues and organs. This requires thousands of genes coded in millions to billions of DNA subunits. Nevertheless, the essential process of evolution by variation and selection remains the same in the test tube, in viruses, and far beyond.

Explaining Order

There are at least three ways to explain the structure of an evolved population of molecular replicators, whether test tube RNA, viral genes, or human genes. The first kind of explanation is a *blow-by-blow* account of their histories: how specific mutations occurred and how they spread. This is impossible without recording all the molecular events, and such a record would in any event be immensely tedious.

The second kind of explanation resorts to a somewhat misleading word: *purpose*. In detail, the molecules simply change haphazardly and replicate selectively. Yet stepping back from the process, one could describe the outcome by imagining that the surviving molecules have changed to "*achieve the goal*" of replication. Why do RNA molecules that evolved under the threat of ribonuclease fold as they do? Because of a long and detailed history, of course, but the idea that "they want to avoid attack and survive to replicate" would predict the same result. The language of purpose makes useful short hand (try discussing human action without it!), but the appearance of purpose need not result from the action of a mind. The RNA example shows this quite neatly.

The third (and often best) kind of explanation—in terms of evolution—says that *order* emerges through the variation and selection of replicators. A molecule folds in a particular way because it resembles ancestors that multipled more successfully (by avoiding attack, etc.), and left descendants including itself. As Richard Dawkins points out, the language of purpose (if used carefully) can be translated into the language of evolution.

Evolution attributes patterns of success to the *elimination* of *unsuccessful* changes. It thus explains a positive as the result of a double negative—an explanation of a sort that seems slightly difficult to grasp. Worse, it explains something *visible* (successful, purposeful entities) in terms of something *invisible* (unsuccessful entities that have vanished). Because only successful beasts have littered the landscape with the bones of their descendants, the malformed failures of the past haven't even left many fossils.

The human mind tends to focus on the visible, seeking positive causes for positive results, an ordering force behind orderly results. Yet through reflection we can see that this great principle has changed our past and will shape our future: *Evolution proceeds by the variation and selection of replicators.*

Evolving Organisms

The history of life is the history of an arms race based on molecular machinery. Today, as this race approaches a new and swifter phase, we need to be sure we understand just how deeply rooted evolution is. In a time when the idea of biological evolution is often slighted in the schools and sometimes attacked, we should remember that the supporting evidence is as solid as rock and as common as cells.

In pages of stone, the Earth itself has recorded the history of life. On lake bottoms and seabed, shells, bones, and silt have piled, layer on layer. Sometimes a shifting current or a geological upheaval has washed layers away; otherwise they have simply deepened. Early layers, buried deep, have been crushed, baked, soaked in mineral waters, and turned to stone.

For centuries, geologists have studied rocks to read Earth's past. Long ago, they found seashells high in the crushed and crumpled rock of mountain ranges. By 1785—seventy-four-years before Darwin's detested book—James Hutton had concluded that seabed mud had been pressed to stone and raised skyward by forces not yet understood. What else could geologists think, unless nature itself had lied?

They saw that fossil bones and shells differed from layer to layer. They saw that shells in layers *here* matched shells in layers *there*, though the layers might lie deep beneath the land between. They named layers (*A,B,C,D*..., or Osagian, Meramecian, Lower Chesterian, Upper

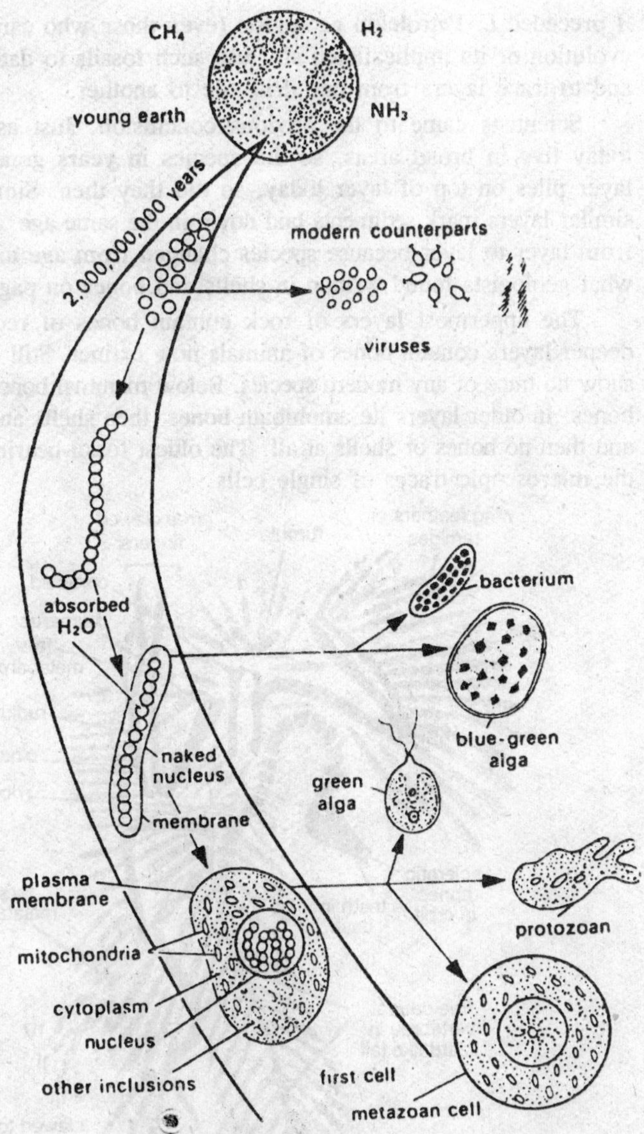

Fig. 5.4. The first organisms.

Chesterian ...), and used characteristic fossils to trace rock layers. The churning of Earth's crust has nowhere left a complete sequence of layers exposed, yet geologists finding *A,B,C,D,E* in one place, *C,D,E,F,G,H,I,J* in another and *J,K,L* somewhere else could see that

A preceded *L*. Petroleum geologists (even those who care nothing for evolution or its implications) still use such fossils to date rock layers and to trace layers from one drill site to another.

Scientists came to the obvious conclusion. Just as sea species today live in broad areas, so did species in years gone by. Just as layer piles on top of layer today, so did they then. Similar shells in similar layers mark sediments laid down in the same age. Shells change from layer to layer because species changed from age to age. This is what geologists found written in shells and bones on pages of stone.

The uppermost layers of rock contain bones of recent animals, deeper layers contain bones of animals now extinct. Still earlier layers show no trace of any modern species. Below mammal bones lie dinosaur bones; in older layers lie amphibian bones, then shells and fish bones, and then no bones or shells at all. The oldest fossil-bearing rocks bear the microscopic traces of single cells.

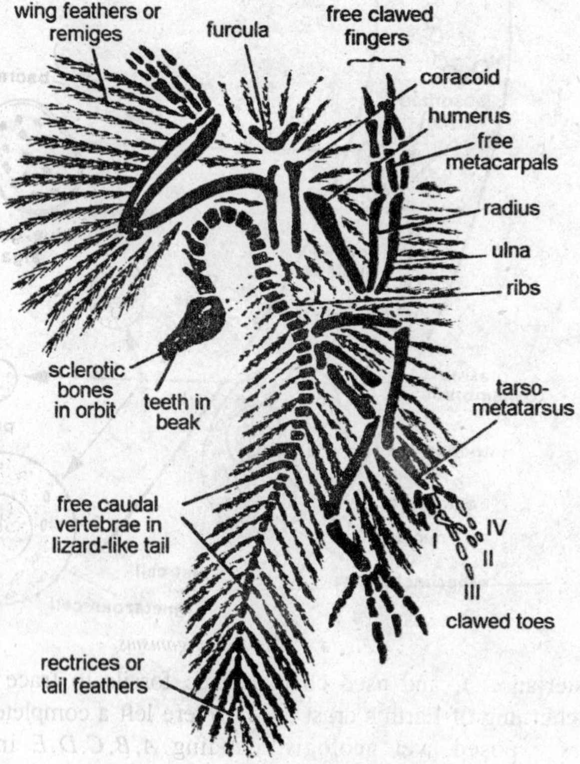

Fig. 5.5. Fossil of Archeopteryx.

Radioactive dating shows these oldest traces to be several billion years old. Cells more complex than bacteria date to little more than one billion years ago. The history of worms, fish, amphibians, rep tiles, and mammals spans hundreds of millions of years. Human-like bones date back several million years. The remains of civilizations date back several thousand.

In three billion years, life evolved from single cells able to soak up chemicals to collections of cells embodying minds able to soak up ideas. Within the last century, technology has evolved from the steam locomotive and electric light to the spaceship and the electronic computer—and computers are already being taught to read and write. With mind and technology, the rate of evolution has jumped a million fold or more.

Another Route Back

The book of stone records the forms of long-dead organisms, yet living cells also carry records, genetic texts only now being read. As with the ideas of geology, the essential ideas of evolution were known before Darwin had set pen to paper.

In lamp-lit temples and monasteries, generations of scribes copied and recopied manuscripts. Sometimes they miscopied words and sentences—whether by accident, by perversity, or by order of the local ruler—and as the manuscripts replicated, aided by these human copying machines, errors accumulated. The worst errors might be caught and removed, and famous passages might survive unchanged, but differences grew.

Ancient books seldom exist in their original versions. The oldest copies are often centuries younger than the lost originals. Nonetheless, from differing copies with differing errors, scholars can recon structure versions closer to the original.

They compare texts. They can trace lines of descent from common ancestors because unique patterns of errors betray copying from a common source. Where all surviving copies agree, scholars can assume that the original copy (or at least the last shared ancestor of the survivors) held the same words. Where survivors differ, scholars study copies that descended separately from a distant ancestor, because areas of agreement then indicate a common origin in the ancestral version.

Genes resemble manuscripts written in a four-letter alphabet. Much as a message can take many forms in ordinary language (restating an idea using entirely different words is no great strain), so different genetic wording can direct the construction of identical protein molecules. Moreover, *protein molecules* with different design details can serve identical functions. A collection of *genes* in a cell is like a whole book, and *genes*—like old manuscripts—have been copied and recopied by inaccurate scribes.

Like scholars studying ancient texts, biologists generally work with modern copies of their material. They compare organisms with similar appearances (lions and tigers, horses and zebras, rats and mice) and find that they give similar answers to the essay questions in their genes and proteins. The more two organisms differ (lions and lizards, humans and sunflowers), the more these answers differ, even among molecular machines serving identical functions. More telling still, similar animals make the same mistakes—all *primates*, for example, lack enzymes for making vitamin C, an omission shared by only two other known mammals, the guinea pig and the fruit bat. This suggests that we primates have copied our genetic answers from a shared source, long ago.

The same principle that shows the lines of descent of ancient texts (and that helps correct their copying errors) thus also reveals the lines of descent of modem life. Indeed, it indicates that all known life shares a common ancestor.

Rise of the Replicators

The *first replicators* on Earth evolved abilities beyond those possible to RNA molecules replicating in test tubes. By the time they reached the bacterial stage, they had developed the "modern" system of using DNA, RNA, and ribosomes to construct protein. Mutations then changed not only the replicating DNA itself, but protein machines and the living structures they build and shape.

Teams of genes shaped ever more elaborate cells, then guided the cellular cooperation that formed complex organisms. Variation and selection favoured teams of genes that shaped beasts with protective skins and hungry mouths, animated by nerve and muscle, guided by eye and brain. As Richard Dawkins puts it, genes built ever more elaborate survival machines to aid their own replication.

When *dog genes* replicate, they often shuffle with those of other dogs that have been selected by people, who then select which puppies to keep and breed. Over the millennia, people have molded wolf-like beasts into greyhounds, toy poodles, dachshunds, and Saint Bernards. By selecting which genes survive, people have reshaped dogs in both body and temperament. Human desires have defined success for dog genes; other pressures have defined success for *wolf genes*.

Mutation and selection of genes has, through long ages, filled the world with grass and trees, with insects, fish, and people. More recently, other things have appeared and multiplied—tools, houses, aircraft, and computers. And like the lifeless RNA molecules, this hardware has *evolved*.

Evolving Technology

As the stone of Earth records the emergence of ever more complex and capable forms of life, so the relics and writings of humanity record the emergence of ever more complex and capable forms of hardware. Our oldest surviving hardware is itself stone, buried with the fossils of our ancestors; our newest hardware orbits overhead.

Consider for a moment the hybrid ancestry of the space shuttle. On its aircraft side, it descends from the aluminum jets of the sixties, which themselves sprang from a line stretching back through the aluminum prop planes of World War II, to the wood-and-cloth biplanes of World War I, to the motorized gliders of the Wright brothers, to toy gliders and kites. On its rocket side, the shuttle traces back to Moon rockets, to military missiles, to last century's artillery rockets ("and the rocket's red glare . . ."), and finally to fireworks and toys. This aircraft/rocket hybrid flies, and by varying components and designs, aerospace engineers will evolve still better ones.

Engineers speak of "*generations*" of technology; Japan's "fifth-generation" computer project shows how swiftly some technologies grow and spawn. Engineers speak of "*hybrids*," of "competing technologies," and of their "proliferation." IBM Director of Research Ralph E. Gomory emphasizes the evolutionary nature of technology, writing that "technology development is much more evolutionary and much less revolutionary or breakthrough-oriented than most people imagine (Indeed, even breakthroughs as important as molecular assemblers will develop through many small steps.) In the quote that heads this chapter, Professor Herbert A. Simon of Carnegie-Mellon University urges us to "think of the design process as involving first the generation of alternatives and then the testing of these alternatives against a whole array of requirements and constraints." Generation and testing of alternatives is synonymous with variation and selection.

Sometimes various alternatives already exist. In "One Highly evolved Toolbox," in *The Next Whole Earth Catalog*, J. Baldwin writes: "Our portable shop has been evolving for about twenty years now. There's nothing really very special about it except that a con turning process of removing obsolete or inadequate tools and replacing them with more suitable ones has resulted in a collection that has become a thing-making system rather than a pile of hardware."

Baldwin uses the term "*evolving*" accurately. Invention and manufacture have for millennia generated variations in tool designs, and Baldwin has winnowed the current crop by competitive selection, keeping those that work best with his other tools to serve his needs. 'Through years of variation and selection, his system evolved—a process he highly recommends. Indeed, he urges that one never try to plan out the purchase of a complete set of tools. Instead, he urges buying the tools one often borrows, tools selected not by theory but by experience.

Technological variations are often deliberate, in the sense that engineers are paid to invent and test. Still, some novelties are sheer accident, like the discovery of a crude form of Teflon in a cylinder supposedly full of tetrafluoroethylene gas: with its valve open, it remained heavy; when it was sawed open, it revealed a strange, waxy solid. Other novelties have come from systematic blundering. Edison tried carbonizing everything from paper to bamboo to spiderwebs when he was seeking a good light-bulb filament. Charles Goodyear messed around in a kitchen for years, trying to convert gummy natural rubber into a durable substance, until at last he chanced to drop sulfurized rubber on a hot stove, performing the first crude vulcanization.

In engineering, enlightened trial and error, not the planning of flawless intellects, has brought most advances; this is why engineers build prototypes. Peters and Waterman in their book *In Search of Excellence* show that the same holds true of advances in corporate products and policies. This is why excellent companies create "an environment and a set of attitudes that encourage experimentation," and why they evolve "*in a very Darwinian way*."

Factories bring order through variation and selection. Crude quality-control systems test and discard faulty parts before assembling products, and sophisticated quality-control systems use statistical methods to track defects to their sources, helping engineers change the manufacturing process to minimize defects. Japanese engineers, building on W. Edwards Deming's work in statistical quality control, have made such variation and selection of industrial processes a pillar of their country's economic success. Assembler-based systems will likewise need to measure results to eliminate flaws.

Quality control is a sort of evolution, aiming not at change but at eliminating harmful variations. But just as *Darwinian evolution* can

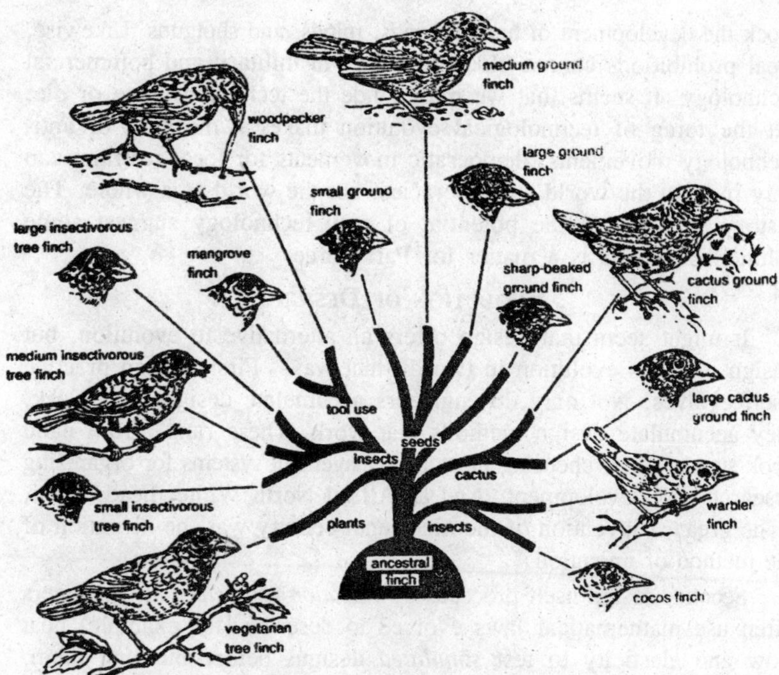

Fig. 5.6. Darwin finches.

preserve and spread favourable mutations, so good quality control systems can help managers and workers to preserve and spread more effective processes, whether they appear by accident or by design.

All this tinkering by engineers and manufacturers prepares products for their ultimate test. Out in the market, endless varieties of wrench, car, sock, and computer compete for the favour of buyers. When informed buyers are free to choose, products that do too little or cost too much eventually fail to be reproduced. As in nature, competitive testing makes yesterday's best competitor into tomorrow's fossil. "*Ecology*" and "*economy*" share more than linguistic roots.

Both in the marketplace and on real and imaginary battlefields, global competition drives organizations to invent, buy, beg, and steal ever more capable technologies. Some organizations compete chiefly to serve people with superior goods, others compete chiefly to intimidate them with superior weapons. The pressures of evolution drive both.

The global technology race has been accelerating for billions of years. The earthworm's blindness could not block the development of sharp-eyed birds. The bird small brain and clumsy wings could not block the development of human hands, minds, and shotguns. Likewise, local prohibitions cannot block advances in military and commercial technology. It seems that we must guide the technology race or die, yet the force of technological evolution makes a mockery of anti-technology movements: democratic movements for local restraint can only restrain the world's democracies, not the world as a whole. The history of life and the potential of new technology suggest some solutions, but this is a matter for Part Three.

EVOLUTION OF DESIGN

It might seem that design offers an alternative to evolution, but design involves evolution in two distinct ways. First, design practice itself evolves. Not only do engineers accumulate designs that work, they accumulate design methods that work. These range from hand book standards for choosing pipes to management systems for organizing research and development. And as Alfred North White- head stated, "The greatest invention of the nineteenth century was the invention of the method of invention."

Second, design itself proceeds by *variation* and *selection*. Engineers often use mathematical laws evolved to describe (for example) heat flow and elasticity to test *simulated* designs before building them. They thus evolve plans through a cycle of design, calculation, criticism, and redesign, avoiding the expense of cutting metal. The creation of designs thus proceeds through a nonmaterial form of evolution.

Hooke's law, for example, describes how metal bends and stretches: deformation is proportional to the applied stress: twice the pull, twice the stretch. Though only roughly correct, it remains fairly accurate until the metal's springiness finally yields to stress. Engineers can use a form of Hooke's law to design a bar of metal that can support a load without bending too far—and then make it just a bit thicker to a flow for inaccuracies in the law and in their design calculations. They can also use a form of Hooke's law to describe the bending and twisting of aircraft wings, tennis rackets, and automobile frames. But simple mathematical equations don't wrap smoothly around such convoluted structures. Engineers have to fit the equations to simpler shapes (to pieces of the design), and then assemble these partial solutions to describe the flexing of the whole. It is a method (called "*finite element analysis*") that typically requires immense calculations, and without computers it would be impractical. With them, it has grown common.

Such simulations extend an ancient trend. We have always imagined consequences, in hope and fear, when we have needed to select a course of action. Simpler mental models (whether inborn or learned) undoubtedly guide animals as well. When based on accurate mental models, thought experiments can replace more costly (or even deadly) physical experiments—a development evolution has favoured. Engineering simulations simply extend this ability to imagine consequences, to make our mistakes in thought rather than deed.

In "One Highly Evolved Toolbox," J. Baldwin discusses how tools and thought mesh in job-shop work: "You begin to build your tool capability into the way you think about making things. As anyone who makes a lot of stuff will tell you, the tools soon become sort of an automatic part of the design process ... But tools can't become part of your design process if you don't know what is available and what the various tools do."

Having a feel for tool capabilities is essential when planning a job- shop project for delivery next Wednesday; it is equally essential when shaping a strategy for handling the breakthroughs of the coming decades. The better our feel for the future's tools, the sounder will be our plans for surviving and prospering.

A craftsman in a job shop can keep tools in plain sight; working with them every day makes them familiar to his eyes, hands, and mind. He gets to know their abilities naturally, and can put this knowledge to immediate creative use. But people—like us—who have to understand the future face a greater challenge, because the future's tools exist now only as ideas and as possibilities implicit in natural law. These tools neither hang on the wall nor impress them selves on the mind through sight and sound and touch—nor will they, until they exist as hardware. In the coming years of preparation only study, imagination, and thought can make their abilities real to the mind.

What are the new Replicators?

History shows us that hardware evolves. *Test tube RNA*, *viruses*, and *dogs* all show how evolution proceeds by the modification and testing of replicators. But hardware (today) cannot reproduce itself so where are the replicators behind the evolution of technology? What are the machine genes?

Of course, we need not actually identify replicators in order to recognize evolution. Darwin described evolution before Mendel discovered genes, and geneticists learned much about heredity before Watson and Crick discovered the structure of DNA. Darwin needed no knowledge of molecular genetics to see that organisms varied and that some left more descendants.

Fig. 5.7. Variety of dogs.

A replicator is a pattern that can get copies of itself made. It may need help; without protein machines to copy it, DNA could not replicate. But by this standard, some *machines* are replicators! Companies often make machines that fall into the hands of a competitor; the competitor then learns their secrets and builds copies. Just as genes "use" protein machines to replicate, so such machines "use" human minds and hands to replicate. With nanocomputers directing assemblers and disassemblers, the replication of hardware could even be automated.

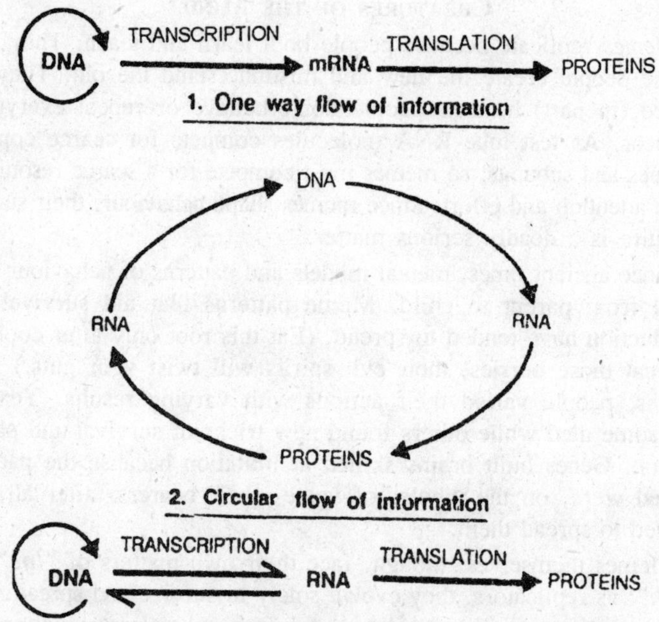

Fig. 5.8. Central dogma.

The human mind, though, is a far subtler engine of imitation than any mere protein machine or assembler. Voice, writing, and drawing can transmit designs from mind to mind before they take form as hardware. The ideas behind methods of design are subtler yet: more abstract than hardware, they replicate and function exclusively in the world of minds and symbol systems.

Where genes have evolved over generations and eons, mental replicators now evolve over days and decades. Like genes, ideas split, combine, and take multiple forms (genes can be transcribed from DNA to RNA and back again; ideas can be translated from language to language). Science cannot yet describe the neural patterns that embody ideas in brains, but anyone can see that ideas mutate, replicate, and compete. Ideas evolve.

Richard Dawkins calls bits of replicating mental patterns "*memes*" (*meme* rhymes with cream). He says "examples of memes are tunes, ideas, catch-phrases, clothes fashions, ways of making pots or of building arches. Just as genes propagate themselves in the gene pool by leaping from body to body (generation to generation) via sperms or eggs, so memes propagate themselves in the meme pool by leaping from brain to brain via a process which, in the broad sense, can be called *imitation*."

Creatures of the Mind

Memes replicate because people both learn and teach. They vary because people create the new and misunderstand the old. They are selected (in part) because people don't believe or repeat everything they hear. As test tube RNA molecules compete for scarce copying machines and subunits, so memes must compete for a scarce resource—human attention and effort. Since memes shape behaviour, their success or failure is a deadly serious matter.

Since ancient times, mental models and patterns of behaviour have passed from parent to child. Meme patterns that aid survival and reproduction have tended to spread. (Eat this root only after cooking; don't eat those berries, their evil spirits will twist your guts.) Year by year, people varied their actions with varying results. Year by year, some died while others found new tricks of survival and passed them on. Genes built brains skilled at imitation because the patterns imitated were, on the whole, of value—their bearers, after all, had survived to spread them.

Memes themselves, though, face their own matters of "*life*" and "*death*": as replicators, they evolve solely to survive and spread. Like viruses, they can replicate without aiding their host's survival or well being. Indeed, the meme for martyrdom-in-a-cause can spread itself through the very act of killing its host.

Genes, like memes, survive by many strategies. Some duck genes have spread themselves by encouraging ducks to pair off to care for their gene-bearing eggs and young. Some duck genes have spread themselves (when in male ducks) by encouraging rape, and some (when in female ducks) by encouraging the planting of eggs in other ducks' nests. Still other genes found in ducks are virus genes, able to spread without making more ducks. Protecting eggs helps the duck species (and the individual duck genes) survive; rape helps one set of duck genes at the expense of others; infection helps viral genes at the expense of duck genes in general. As Richard Dawkins points out, genes "*care*" only about their own replication: they appear selfish.

But selfish motives can encourage cooperation. People seeking money and recognition for themselves cooperate to build corporations that serve other people wants. Selfish genes cooperate to build organisms that themselves often cooperate. Even so, to imagine that genes *automatically* serve some greater good (—of their chromosome? —their cell? —their body? —their species?) is to mistake a common effect for an underlying cause. To ignore the selfishness of replicators is to be lulled by a dangerous illusion. Some genes in cells are out-and-out parasites. Like herpes genes inserted in human chromosomes, they exploit cells and harm their hosts. Yet if genes can be parasites, why not memes as well?

In *The Extended Phenotype*, Richard Dawkins describes a worm that parasitizes bees and completes its life cycle in water. It gets from bee to water by making the host bee dive to its death. Similarly, ant brainworms must enter a sheep to complete their life cycle. To accomplish this, they burrow into the host ant's brain, somehow causing changes that make the ant "*want*" to climb to the top of a grass stem and wait, eventually to be eaten by a sheep.

As worms enter other organisms and use them to survive and replicate, so do memes. Indeed, the *absence* of memes exploiting people for their own selfish ends would be amazing, a sign of some powerful—indeed, nearly perfect—mental immune system. But parasitic memes clearly do exist. Just as viruses evolve to stimulate cells to make viruses, so rumors evolve to sound plausible and juicy, stimulating repetition. Ask not whether a rumor is true, ask instead how it spreads Experience shows that ideas evolved to be successful replicators need have little to do with the truth.

At best, chain letters, spurious rumors, fashionable lunacies, and other mental parasites harm people by wasting their time. At worst, they implant deadly misconceptions. These meme systems exploit human ignorance and vulnerability. Spreading them is like having a cold and sneezing on a friend. Though some memes act much like viruses, infectiousness isn't necessarily bad (think of an infectious grin, or infectious good nature). If a package of ideas has merit, then its infectiousness simply increases its merit—and indeed, the best ethical teachings also teach us to teach ethics. Good publications may entertain, enrich understanding, aid judgment—and advertise gift subscriptions. Spreading useful meme systems is like offering useful seeds to a friend with a garden.

Selecting Ideas

Parasites have forced organisms to evolve immune systems, such as the enzymes that bacteria use to cut up invading viruses, or the moving white blood cells our bodies use to destroy bacteria. Parasitic memes have forced minds down a similar path, evolving meme systems that serve as mental immune systems.

The oldest and simplest mental immune system simply commands "*believe the old, reject the new.*" Something like this system generally kept tribes from abandoning old, tested ways in favour of wild new notions—such as the notion that obeying alleged ghostly orders to destroy all the tribe's cattle and grain would somehow bring forth a miraculous abundance of food and armies of ancestors to drive out foreigners. (This meme package infected the Xhosa people of southern Africa in 1856; by the next year 68,000 had died, chiefly of starvation.)

Your body's immune system follows a similar rule: it generally accepts all the cell types present in early life and rejects new types, such as potential cancer cells and invading bacteria, as foreign and dangerous. This simple reject-the-new system once worked well, yet in this era of organ transplantation it can kill. Similarly, in an era when science and technology regularly present facts that are both new and trustworthy, a rigid mental immune system becomes a dangerous handicap.

For all its shortcomings, though, the reject-the-new principle is simple and offers real advantages. Tradition holds much that is tried and true (or if not true, then at least workable). Change is risky: just as most mutations are bad, so most new ideas are wrong. Even reason can be dangerous: if a tradition links sound practices to a fear of ghosts, then overconfident rational thought may throw out the good with the bogus. Unfortunately, traditions evolved to be good may have less appeal than ideas evolved to *sound* good—when first questioned, the soundest tradition may be displaced by worse ideas that better appeal to the rational mind.

Yet memes that seal the mind against new ideas protect themselves in a suspiciously self-serving way. While protecting valuable traditions from clumsy editing, they may also shield parasitic claptrap from the test of truth. In times of swift change they can make minds dangerously rigid.

Much of the history of philosophy and science may be seen as a search for better mental immune systems, for better ways to reject the false, the worthless, and the damaging. The best systems respect tradition, yet encourage experiment. They suggest standards for judging memes, helping the mind distinguish between parasites and tools.

The principles of evolution provide a way to view change, whether in molecules, organisms, technologies, minds, or cultures. The same basic questions keep arising: What are the replicators? How do they vary? What determines their success? How do they defend against invaders? These questions will arise again when we consider the consequences of the assembler revolution, and yet again when we consider how society might deal with those consequences. The deep-rooted principles of evolutionary change will shape the development of nanotechnology, even as the distinction between hardware and life begins to blur. These principles show much about what we can and cannot hope to achieve, and they can help us focus our efforts to shape the future. They also tell us much about what we can and cannot foresee, because they guide the evolution not only of hardware, but of knowledge itself.

6

Limits to Growth

Aircraft, spacecraft, nuclear power, and computers have already been developed. In the near future we will be in a position to develop assemblers, replicators, automated engineering, cheap spaceflight, cell re pair machines, and much more. This series of breakthroughs may suggest that the technology race will advance without limit. In this view, we will break through all conceivable barriers, rushing off into the infinite unknown–but this view seems false.

The *laws of nature* and the conditions of the world will limit what we do. Without limits, the future would be wholly unknown, a form less thing making a mockery of our efforts to think and plan. With limits, the future is still a turbulent uncertainty, but it is forced to flow within certain bounds.

From *natural limits*, we learn something about the problems and opportunities we face. *Limits* define the boundaries of the possible, telling us what resources we can use, how fast our spacecraft will fly, and what our *nanomachines* will and won't be able to do. Discussing limits is risky: we can be more sure that something is possible than that it is not. Engineers can make do with approximations and special cases. And given tools, materials, and time, they can demonstrate possibilities directly. Even when doing exploratory de sign, they can stay well within the realm of the possible by staying well away from the limits. Scientists, in contrast, cannot prove a general theory–and every general claim of impossibility is itself a sort of general theory. No specific experiment (someplace, sometime) can prove something to be impossible (everywhere, forever). Neither can any number of specific experiments.

Still, *general scientific laws* do describe limits to the possible. Although scientists cannot *prove* a general law, they have evolved our best available picture of how the universe works. And even if exotic experiments and elegant mathematics again transform our concept of physical law, few engineering limits will budge. Relativity didn't affect automobile designs.

The mere existence of ultimate limits doesn't mean that they are about to choke us, yet many people have been drawn to the idea that limits will end growth soon. This notion simplifies their picture of the future by leaving out the strange new developments that growth will bring. Other people favour the vaguer notion of limitless growth—a notion that blurs their picture of the future by suggesting that it will be utterly incomprehensible.

People who confuse science with technology tend to become con fused about limits. As software engineer Mark S. Miller points out, they imagine that new knowledge always means new know-how; some even imagine that knowing *everything* would let us do *anything*. Advances in technology do indeed bring new know-how, opening new possibilities. But advances in basic science simply redraw our map of ultimate limits; this often shows new impossibilities. Einstein's discoveries, for example, showed that nothing can catch up with a fleeing light ray.

Structure of the Vacuum

Is the speed of light a real limit? People once spoke of a "*sound barrier*" that some believed would stop an airplane from passing the speed of sound. Then at Edwards Air Force Base in 1947, Chuck Yeager split the October sky with a sonic boom. Today, some people speak of a "*light barrier*," and ask whether it, too, may fall.

Unfortunately for science fiction writers, this parallel is superficial. No one could ever maintain that the speed of sound was a true physical limit. Meteors and bullets exceeded it daily, and even cracking whips cracked the "*sound barrier.*" But no one has seen anything move faster than light. Distant spots seen by radio telescopes sometimes *appear* to move faster, but simple tricks of perspective easily explain how this can be. Hypothetical particles called "*tachyons*" would move faster than light, if they were to exist—but none has been found, and current theory doesn't predict them. Experimenters have pushed *protons* to over 99.9995 percent of the speed of light, with results that match Einstein's predictions perfectly. When pushed ever faster, a particle's speed creeps closer to the speed of light, while its energy (mass) grows without bound.

Light Barrier

On Earth, a person can walk or sail only to a certain distance, but no mysterious edge or barrier suddenly blocks travel. The Earth is simply round. The speed limit in space no more implies a "*light barrier*" than the distance limit on Earth implies a wall. Space itself—the vacuum that holds all energy and matter—has properties. One of these is its geometry, which can be described by regarding time as a special dimension. This geometry makes the speed of light recede before an accelerating spaceship much as the horizon recedes before a moving sea ship: the speed of light, like the horizon, is always equally remote in all directions. But the analogy dies here—this similarity has nothing to do with the curvature of space. It is enough to remember that the limiting speed is nothing so crude or so breakable as a "*light barrier.*" Objects can always go faster, just no faster than light.

Gravity

People have long dreamed of *gravity control*. In the 1962 edition of *Profiles of the Future*, Arthur C. Clarke wrote that "Of all the forces, gravity is the most mysterious and the most implacable," and then went on to suggest that we will someday develop convenient devices for controlling gravity. Yet is gravity really so mysterious? In the general theory of relativity, Einstein described gravity as curvature in the space-time structure of the vacuum. The mathematics describing this is elegant and precise, and it makes predictions that have passed every test yet contrived.

Gravity is neither more nor less implacable than other forces. No one can make a boulder lose its gravity, but neither can anyone make an electron lose its electric charge or a current its magnetic field. We control electric and magnetic fields by moving the particles that create them; we can control gravitational fields similarly, by moving ordinary masses. It seems that we cannot learn the secret of gravity control because we already have it.

A child with a small magnet can lift a nail, using a magnetic field to overwhelm Earth's gravitational pull. But unfortunately for eager gravitational engineers, using gravity to lift a nail requires a tremendous mass. Hanging Venus just over your head would almost do the job—until it fell on you.

Engineers stir up electromagnetic waves by shaking electric charges back and forth in an antenna; one can stir up gravity waves by shaking a rock in the air. But again, the gravitational effect is weak. Though a one-kilowatt radio station is nothing extraordinary, all the shaking and spinning of all the masses in the solar system put together fails to radiate as much as a kilowatt of gravity waves.

We understand gravity well enough; it simply isn't much use in building machines much lighter than the Moon. But devices using large masses do work. A *hydroelectric* dam is part of a gravity machine (the other part being the Earth) that extracts energy from falling mass. Machines using black holes will be able to extract energy from falling mass with over fifty percent efficiency, based on $E = mc^2$. Lowering a single bucket of water into a black hole would yield as much energy as pouring several trillion buckets of water through the generators of a kilometer-high dam.

Because the *laws of gravity* describe how the vacuum curves, they also apply to science-fiction style space warps." It seems that tunnels from one point in space to another would be unstable, even if they could be created in the first place. This prevents future spaceships from reaching distant points faster than light by taking a shortcut around the intervening space, and this limit to travel in turn sets a limit to growth.

Einstein's laws seem to give an accurate description of the overall geometry of the vacuum. If so, then the limits that result seem inescapable: you can get rid of almost anything, but not the vacuum itself.

Other laws and limits seem inescapable for similar reasons. In fact, physicists have increasingly come to regard all of physical law in terms of the structure of the vacuum. Gravity waves are a certain kind of ripple in the vacuum; black holes are a certain kind of kink. Likewise, radio waves are another kind of ripple in the vacuum, and elementary particles are other, very different kinds of kinks (which in some theories resemble tiny, vibrating strings). In this view, there is only one substance in the universe—the vacuum—but one that takes on a variety of forms, including those patterns of particles that we call "solid matter." This view suggests the inescapable quality of natural law. If a single substance fills the universe, is the universe, then its properties limit all that we can do.

The strangeness of modern physics, however, leads many people to distrust it. The revolutions that brought quantum mechanics and relativity gave rise to talk of "*the uncertainty principle*," "the wave nature of matter," "matter being energy," and "curved space-time."

An air of paradox surrounds these ideas and thus physics itself. It is understandable that new technologies should seem odd to us, but why should the ancient and immutable laws of nature turn out to be bizarre and shocking?

Our brains and languages evolved to deal with things vastly larger than atoms, moving at a tiny fraction of the speed of light. They do a tolerable job of this, though it took people centuries to learn to describe the motion of a falling rock. But we have now stretched our knowledge far beyond the ancient world of the senses. We have found things (matter waves, curved space) that seem bizarre—and some that are simply beyond our ability to visualize. But "bizarre" does not mean mysterious and unpredictable. Mathematics and experiments still work, letting scientists vary and select theories, evolving them to fit even a peculiar reality. Human minds have proved remarkably flexible, but it is no great surprise to find that we cannot always visualize the invisible.

Part of the reason that physics seems so strange is that people crave oddities, and tend to spread memes that describe things as odd. Some people favour ideas that coat the world in layers of ghost stuff and fill it with grade-B mysteries. Naturally, they favour and spread memes that make matter seem immaterial and quantum mechanics seem like a branch of psychology.

Energy

Relativity, it is said, reveals that *matter* (that plain old stuff that people think they understand) is really *energy* (that subtle, mysterious stuff that makes things happen). This encourages a smiling vagueness about the mystery of the universe. It might be clearer to say that relativity reveals *energy* to be a form of *matter*: in all its forms, energy has mass. Indeed, light sails work on this principle, through the impact of mass on a surface. Light even comes packaged in particles.

Consider also the Heisenberg uncertainty principle, and the related fact that "the observer always affects the observed." The uncertainty principle is intrinsic to the mathematics describing ordinary matter (giving atoms their very size), but the associated "effect of the observer" has been presented in some popular books as a magical influence of consciousness on the world. In fact, the core idea is more prosaic. Imagine looking at a dust mote in a light beam: When you observe the reflected light, you certainly affect it—your eye absorbs it. Likewise, the light (with its mass) affects the dust mote: it bounces off the mote, exerting a force. The result is not an effect of your mind on the dust, but of the light on the dust. Though quantum measurement has peculiarities far more subtle than this, none involves the mind reaching out to change reality.

Twin Paradox

Finally, consider the "*twin paradox.*" Relativity predicts that, if one of a pair of twins flies to another star and back at near the speed of light, the traveling twin will be younger than the stay-at-home twin. Indeed, measurements with accurate clocks demonstrate the time-slowing effect of rapid motion. But this is not a 'paradox"; it is a simple fact of nature.

Will Physics Again Be Upended?

In 1894 the eminent physicist Albert A. Michelson stated: "The more important fundamental laws and facts of physical science have all been discovered, and these are now so firmly established that the possibility of their ever being supplanted in consequence of new discoveries is exceedingly remote ... Our future discoveries must be looked for in the sixth place of decimals."

But in 1895, Roentgen discovered X rays. In 1896, Becquerel discovered radioactivity. In 1897, Thomson discovered the electron. In 1905, Einstein formulated the special theory of relativity (and thus explained Michelson's own 1887 observations regarding the speed of light). In 1905, Einstein also presented the photon theory of light. In 1911, Rutherford discovered the atomic nucleus. In 1915, Einstein formulated the general theory of relativity. In 1924-30, de Broglie, Heisenberg, Bohr, Pauli, and Dirac developed the foundations of quantum mechanics. In 1929, Hubble announced evidence for the expansion of the universe. In 1931, Michelson died.

Michelson had made a memorable mistake. People still point to his statement and what followed to support the view that we shouldn't (ever?) claim any firm understanding of natural law, or of the limits to the possible. After all, if Michelson was so sure and yet so wrong, shouldn't we fear repeating his mistake? The great revolution in physics has led some people to conclude that science will hold end less important surprises—even surprises important to engineers. But are we likely to encounter such an important upheaval again?

Perhaps not. The content of quantum mechanics was a surprise, yet before it appeared, physics was obviously and grossly incomplete. Before quantum mechanics, you could have walked up to any scientist, grinned maliciously, rapped on his desk and asked, "What holds this thing together? Why is it brown and solid, while air is transparent and gaseous?" Your victim might have said something vague about atoms and their arrangements, but when you pressed for an explanation, you would at best have gotten an answer like "Who knows? Physics can't explain matter yet!" Hindsight is all too easy, yet in a world made of matter, inhabited by material people using material tools, this ignorance of the nature of matter was a gap in human knowledge that Michelson should perhaps have noted. It was a gap not in "the sixth place of decimals" but in the first.

It is also worth observing the extent to which Michelson was *right*. The laws of which he spoke included *Newton's laws* of *gravity* and *motion*, and *Maxwell's laws* of *electromagnetism*. And indeed, under conditions common in engineering, these laws have been modified only "in the sixth place of decimals." *Einstein's laws* of *gravity* and motion match Newton's laws closely except under extreme conditions of gravitation and speed; the *quantum electrodynamic laws* of Feynman, Schwinger, and Tomonaga match Maxwell's laws closely except under extreme conditions of size and energy.

Further revolutions no doubt lurk around the edges of these theories. But those edges seem far from the world of living things and the machines we build. The revolutions of relativity and quantum mechanics changed our knowledge about matter and energy, but matter and energy themselves remained unchanged—they are real and care nothing for our theories. Physicists now use a single set of laws to describe how atomic nuclei and electrons interact in atoms, molecules, molecular machines, living things, planets, and stars. These laws are not yet completely general; the quest for a unified theory of all physics continues. But as physicist Stephen W. Hawking states, "at the moment we have a number of partial laws which govern the behaviour of the universe under all but the most extreme conditions." And by an engineer's standards, those conditions are extra ordinary extreme.

Physicists regularly announce new particles observed in the debris from extremely energetic particle collisions, but you cannot buy one of these new particles in a box. And this is important to recognize, because if a particle cannot be kept, then it cannot serve as a component of a stable machine. Boxes and their contents are made of electrons and nuclei. Nuclei, in turn, are made of *protons* and *neutrons*. Hydrogen atoms have single protons as their nuclei; lead atoms have eighty-two protons and over a hundred neutrons. Isolated neutrons disintegrate in a few minutes. Few other stable particles are known: photons—the particles of light—are useful and can be trapped for a

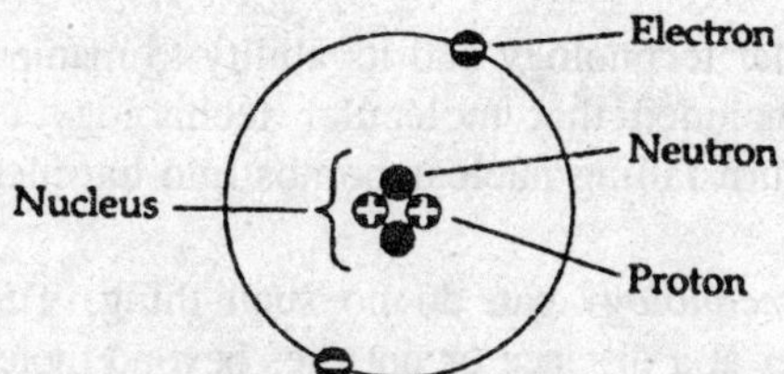

Fig. 6.1. Structure of atom.

time; neutrinos are almost undetectable and cannot even be trapped. These particles (save the photon) have matching antiparticles. *All other known particles disintegrate in a few millionths of a second or less.* Thus, the only known building blocks for hardware are electrons and nuclei (or their antiparticles, for special *isolated* applications); these building blocks ordinarily combine to form atoms and molecules.

Yet despite the power of modern physics, our knowledge still has obvious holes. The shaky state of elementary particle theory leaves some limits uncertain. We may find new stable, "*boxable*" particles, such as magnetic monopoles or free quarks; if so, they will doubtless have uses. We may even find a new long-range field or form of radiation, though this seems increasingly unlikely. Finally, some new way of smashing particles together may improve our ability to convert known particles into other known particles.

But in general, complex hardware will require complex, stable patterns of particles. Outside the environment of a collapsed star, this means patterns of atoms, which are well described by relativistic quantum mechanics. The frontiers of physics have moved on. On a theoretical level, physicists seek a unified description of the interactions of all possible particles, even the most short-lived particles. On an experimental level, they study the patterns of subatomic debris created by high-energy collisions in particle accelerators. So long as no new, stable, and useful particle comes out of such a collision—or turns up as the residue of past cosmic upheavals—atoms will remain the sole building blocks of stable hardware. And engineering will remain a game that is played with already known pieces according to already known rules. New particles would add pieces, not eliminate rules.

Limits to Hardware

Molecular machinery is the end of the road where miniaturization is concerned, is doubtful. The idea that molecular machinery might be a step toward a smaller "*nuclear machinery*" seems natural enough. One young man (a graduate student in economics at Columbia University) heard of molecular technology and its ability to manipulate atoms, and immediately concluded that molecular technology could do almost anything, even turn falling nuclear bombs into harmless lead bricks at a distance.

Molecular technology can do no such thing. Turning plutonium into lead (whether at a distance or not) lies beyond molecular technology for the same reason that turning lead into gold lay beyond an alchemist chemistry. Molecular forces have little effect on the atomic nucleus. The nucleus holds over 99.9 percent of an atom's mass and occupies about 1/1,000,000,000,0000,000 of its volume. Compared to the nucleus, the rest of an atom (an *electron cloud*) is less than airy fluff. Trying to change a nucleus by poking at it with a molecule is even more futile than trying to flatten a steel ball bearing by waving a ball of cotton candy at it. Molecular technology can sort and rearrange atoms, but it cannot reach into a nucleus to change an atom's type.

Nanomachines may be no help in *building* nuclear-scale machines, but could such machines even *exist*? Apparently not, at least under any conditions we can create in a laboratory. Machines must have a number of parts in close contact, but close-packed nuclei repel each other with ferocity. When splitting nuclei blasted Hiroshima, most of the energy was released by the violent electrostatic repulsion of the freshly split halves. The well-known difficulty of nuclear fusion stems from this same problem of nuclear repulsion.

In addition to splitting or fusing, nuclei can be made to emit or absorb various types of radiation. In one technique, they are made to gyrate in ways that yield useful information, letting doctors make medical images based on nuclear magnetic resonance. But all these phenomena rely only on the properties of well-separated nuclei. Isolated nuclei are too simple to act as machines or electronic circuits. Nuclei can be forced close together, but only under the immense pressures found inside a collapsed star. Doing engineering there would present substantial difficulties, even if a collapsed star were handy.

This returns us to the basic question, What can we accomplish by properly arranging atoms? Some limits already seem clear. The strongest materials possible will have roughly ten times the strength of today's strongest steel wire. (The strongest material for making a cable appears to be carbyne, a form of carbon having atoms arranged in straight chains.) It seems that the vibrations of heat will, at ordinary pressures, tear apart the most refractory solids at temperatures around four thousand degrees Celsius (roughly fifteen hundred degrees cooler than the Sun's surface). These brute properties of matter—strength and heat resistance—cannot be greatly improved through complex, clever arrangements of atoms. The best arrangements seem likely to be fairly simple and regular. Other fairly simple goals include transmitting heat, insulating against heat, transmitting electricity, insulating against electricity, transmitting light, reflecting light, and absorbing light.

With some goals, pursuit of perfection will lead to simple designs; with others, it will lead to design problems beyond any hope of solving. Designing the best possible switching component for a computer may prove fairly easy; designing the best possible computer will be vastly more complex. Indeed, what we consider to be "*the best possible*" will depend on many factors, including the costs of matter, energy and time—and on what we want to compute. In any engineering project, what we call "*better*" depends on indefinitely many factors, including ill-defined and changing human wants. What is more, even where 'better" is well defined, the cost of seeking the final increment of improvement that separates the best from the merely excellent may not be worth paying. We can ignore all such issues of complexity and design cost, though, when considering whether limits actually exist.

To define a limit, one must choose a direction, a scale of quality. With that direction defining "*better*," there will definitely be a best. The arrangement of atoms determines the properties of hardware, and according to quantum mechanics, the number of possible arrangements is finite—more than just astronomically large, yet not infinite. It follows mathematically that, given a clear goal, some one of these arrangements must be best, or tied for best. As in chess, the limited number of pieces and spaces limits the arrangements and hence the possibilities. In both chess and engineering, though, the variety possible within those limits is inexhaustible.

Just knowing the fundamental laws of matter isn't enough to tell us exactly where all the limits lie. We still must face the complexities of design. Our knowledge of some limits remains loose: "We know only that the limit lies between here (a few paces away) and there (that spot near the horizon)." Assemblers will open the way to the limits, wherever they are, and automated engineering systems will speed progress along the road. The absolute best will often prove elusive, but the runners-up will often be nearly as good. As we approach genuine limits, our abilities will, in ever more areas of technology, cease growing. Advances in these fields will stop not merely for a decade or a century, but permanently.

Some may balk at the word "*permanently*," thinking "No improvements in a thousand years? In a *million* years? This must be an overstatement." Yet where we reach true physical limits, we will go no further. The rules of the game are built into the structure of the vacuum, into the structure of the universe. No rearranging of atoms, no clashing of particles, no legislation or chanting or stomping will move natural limits one whit. We may misjudge the limits today, but wherever the real limits lie, there they will remain.

This look at natural law shows limits to the quality of things. But we also face limits to *quantity*, set not only by natural law but by the way that matter and energy are arranged in the universe as we hap pen to find it. The authors of *The Limits to Growth*, like so many others, attempted to describe these limits without first examining the limits to technology. This gave misleading results.

Entropy: A Limit to Energy Use

Recently, some authors have described the accumulation of waste heat and disorder as ultimate limits to human activity. In *The Lean Years—Politics in the Age of Scarcity*, Richard Barnet writes:

"It is ironical that the rediscovery of limits coincides with two of the most audacious technological feats in human history. One is genetic engineering, the sudden glimpse of a power to shape the very stuff of life. The other is the colonization of space. These breakthroughs encourage fantasies of power, but they do not break the ecological straightjacket known as the Second Law of Thermodynamics: Ever greater consumption of energy produces ever greater quantities of heat, which never disappear, but must be counted as a permanent energy cost. Since accumulation of heat can cause ecological catastrophe, these costs limit man's adventure in space as surely as on earth."

Jeremy Rifkin (with Ted Howard) has written an entire book on thermodynamic limits and the future of humanity, titled *Entropy: A New World View*.

Entropy is a standard scientific measure of waste heat and disorder. Whenever activities consume useful energy, they produce entropy; the entropy of the world therefore, increases steadily and irreversibly. Ultimately, the dissipation of useful energy will destroy the basis of life. As Rifkin says, this idea may seem too depressing to consider, but he argues that we must face the terrible facts about entropy, humanity, and the Earth. But are these facts so terrible?

Barnet writes that accumulating heat is a permanent energy cost, limiting human action. Rifkin states that "pollution is the sum total of all of the available energy in the world that has been transformed into unavailable energy." This unavailable energy is chiefly low-temperature waste heat, the sort that makes television sets get warm. But does heat really accumulate, as Barnet fears? If so, then the Earth must be growing steadily hotter, minute by minute and year by year. We should be roasting now, if our ancestors weren't frozen solid. Somehow, though, continents manage to get cold at night, and colder yet during the winter. During ice ages, the whole Earth cools off.

Rifkin takes another tack. He states that "the fixed endowment of terrestrial matter that makes up the earth's crust is constantly dissipating. Mountains are wearing down and topsoil is being blown away with each passing second." By "*blown away*" Rifkin doesn't mean blown into space or blown out of existence; he just means that the mountain's atoms have become jumbled together with others. Yet this process, he argues, means our doom. The jumbling of atoms makes them "*unavailable matter*," as a consequence of the "*fourth law of thermodynamics*," propounded by economist Nicholas Georgescu-Roegen: "In a closed system, the material entropy must ultimately reach a maximum," or (equivalently) that "*unavailable matter cannot be recycled.*" Rifkin declares that the Earth is a closed system, exchanging energy but not matter with its surroundings, and that therefore "here on earth material entropy is continually increasing and must ultimately reach a maximum," making Earth's life falter and die.

A grim situation indeed—the Earth has been degenerating for billions of years. Surely the end must be near!

But can this really be true? As life developed, it brought more order to Earth, not less; the formation of ore deposits did the same. The idea that Earth has degenerated seems peculiar at best (but then, Rifkin thinks evolution is bunk). Besides, since matter and energy are essentially the same, how can a valid law single out something called "*material entropy*" in the first place?

Rifkin presents perfume spreading from a bottle into the air in a room as an example of "*dissipating matter*," of material entropy increasing—of matter becoming "*unavailable.*" The spread of salt into water in a bottle will serve equally well. Consider, then, a test of the "*fourth law of thermodynamics*" in the Salt-Water Bottle Experiment: Imagine a bottle having a bottom with a partition, dividing it into two basins. In one sits salt, in the other sits water. A cork plugs the bottle's neck: this closes the system and makes the so-called fourth law of thermodynamics apply. The bottle's contents are in an organized state: their material entropy is not at a maximum—yet.

Now, pick up the bottle and shake it. Slosh the water into the other basin, swirl it around, dissolve the salt, increase the entropy—go wild! In such a closed system, the "*fourth law of thermodynamics*" says that this increase in the material entropy should be permanent. All of Rifkin's alarums about the steady, inevitable increase of Earth's entropy rest on this principle. To see if there is any basis for Rifkin's new worldview, take the bottle and tip it, draining the salty water into one basin. This should make no difference, since the system remains closed. Now set the bottle upright, placing the saltwater side in sunlight and the empty side in shade. Light shines in and heat leaks out, but the system remains as closed as the Earth itself. But watch—the sunlight evaporates water, which condenses in the shade! Fresh water slowly fills the empty basin, *leaving the salt behind*.

Rifkin himself states that "*in science*, only one uncompromising exception is enough to invalidate a law." This thought experiment, which mimics how natural salt deposits have formed on Earth, invalid dates the law on which he founds his Whole book. So do plants. Sunlight brings energy from space; heat radiated back into space carries away entropy (of which there is only one kind). Therefore, entropy can decline in a closed system and flowers can bloom on Earth for age upon age.

Rifkin is right in saying that "it's possible to reverse the entropy process in an isolated time and place, but Only by using up energy in the process and thus increasing the over all entropy of the environment." But both Rifkin and Barnet make the same mistake: when they write of the environment, they imply the Earth—but the law applies to the environment as a whole, and that whole is the *universe*. In effect, Rifkin and Barnet ignore both the light of the Sun and the cold black of the night sky.

According to Rifkin, his ideas destroy the notion of history as progress, transcending the modern world He calls for sacrifice, stating that "no Third World nation should harbor hopes that it can ever reach the material abundance that ha existed in America." He fears panic and bloodshed. Rifkin finishes by informing us that "the Entropy Law answers the central question that every culture through out history has grappled with: How should human beings behave in the world?" His answer? "The ultimate moral imperative, then, is to waste as little energy as possible."

This would seem to mean that we must save as much energy as possible, seeking to eliminate waste. But "hat is the greatest nearby energy waster? Why, the Sun, of course—it Wastes energy trillions of times faster than we humans do. If taken seriously, it seems that Rifkin's ultimate moral imperative therefore urges: "*Put out the Sun*!"

This silly consequence should have tipped Rifkin off. He and many others hold views that smack of a Pre-Copernican arrogance: they presume that the Earth is the whole World and that what people do is automatically of cosmic importance.

There is a *genuine entropy law*, of course: the second law of thermodynamics. Unlike the bogus "*fourth law*," it is described in text books and used by engineers. It will indeed limit what we do. Human activity will generate heat, and Earth's limited ability to radiate heat will set a firm limit to the amount of Earth-based industrial activity. Likewise, we will need winglike panels to radiate waste heat from our starships. Finally, the entropy law Will at the far end of an immensity of time—bring the downfall of the universe as we know it, limiting the lifespan of life itself.

Why flog the carcass of Rifkin's *Entropy*? Simply because today's information systems often present even stillborn ideas as if they were alive. By encouraging false hopes, false fears, and misguided action, these ideas can waste the efforts of people actively concerned about long-range world problems.

Among those whose praise appears on the back cover of Rifkin's book ("an inspiring work," "brilliant work," "earthshaking," "should be taken to heart") are a Princeton professor, a talk-show host, and two United States senators. A seminar at MIT featured Rifkin's book.

Georgescu-Roegen, inventor of the "fourth law of thermodynamics," has extensive credentials—as a social scientist. The entropy threat is an example of blatant nonsense, yet its inventors and promoters aren't laughed off the public stage. Imagine a thousand, a million similar distortions—some subtle, some brazen, but all warping the public's understanding of the world. Now imagine a group of democratic nations suffering from an infestation of such memes while attempting to cope with an era of accelerating technological revolution. We have a real problem. To make our survival more likely, we will need better ways to weed our memes, to make room for sound understanding to grow.

Limits to Resources

Natural law limits the quality of technology, but within these limits we will use replicating assemblers to produce superior spacecraft. Today Earth has begun to seem small, arousing concerns that we may deplete its resources. Yet the energy we use totals less than 1/10,000 of the solar energy striking Earth; we worry not about the supply of energy as such, but about the supply of convenient gas and oil. Our mines barely scratch the surface of the globe; we worry not about the sheer quantity of resources, but about their convenience and cost. When we develop pollution-free nanomachines to gather solar energy and resources, Earth will be able to support a civilization far larger and wealthier than any yet seen, yet suffer less harm than we inflict today. The potential of Earth makes the resources we now use seem insignificant by comparison.

Yet Earth is but a speck. The asteroidal debris left over from the formation of the planets will provide materials enough to build a thousand times Earth's land area. The Sun floods the solar system with a billion times the power that reaches Earth. The resources of the solar system are truly vast, making the resources of Earth seem insignificant by comparison.

Yet the solar system is but a speck. The stars that crowd the night sky are suns, and the human eye can see only the closest. Our galaxy holds a hundred billion suns, and many no doubt pour their light on dead planets and asteroids awaiting the touch of life. The resources of our galaxy make even our solar system seem insignificant by comparison.

Yet our galaxy is but a speck. Light older than our species shows galaxies beyond ours. The visible universe holds a hundred billion galaxies, each a swarm of billions of suns. The resources of the visible universe make even our galaxy seem insignificant by comparison.

With this we reach the limits of knowledge, if not of resources. The solar system seems answer enough to Earth's limits—and if the rest of the universe remains unclaimed by others, then our prospects for expansion boggle the mind several times over. Does this mean that replicating assemblers and cheap spaceflight will end our resource worries? In a sense, opening space will burst our limits to growth, since there is no end to the universe. Nevertheless, Malthus was essentially right.

Malthus

In his 1798 *Essay on the Principle of Population*, Thomas Robert Malthus, an English clergyman, presented the ancestor of all modern limits-to-growth arguments. He noted that freely growing populations tend to double periodically, thus expanding exponentially. This makes sense: since all organisms are descended from successful replicators, they tend to replicate when given a chance. For the sake of argument, Malthus assumed that resources—the food supply—could increase by a fixed amount per year (a process called *linear growth*, since it plots as a line on a graph). Since mathematics shows that any fixed rate of *exponential growth* will eventually outstrip any fixed rate of linear growth, Malthus argued that population growth, if unchecked, would eventually outrun food production.

Authors have repeated variations on this idea ever since, in books like *The Population Bomb and Famine—1975*!, yet food production has kept pace with population. Outside Africa, it has even pulled ahead. Was Malthus wrong?

Not fundamentally: he was wrong chiefly about timing and details. Growth on Earth does face limits, since Earth has limited room, whether for farming or anything else. Malthus failed to predict when limits would pinch us chiefly because he failed to anticipate breakthroughs in farm equipment, crop genetics, and fertilizers.

Some people now note that exponential growth will overrun the fixed stock of Earth's resources, a simpler argument than the one Malthus made. Though space technology will break this limit, it will not break all limits. Even if the universe were infinitely large, we still could not travel infinitely fast. The laws of nature will limit the rate of growth: Earth's life will spread no faster than light.

Steady expansion will open new resources at a rate that will increase as the frontier spreads deeper and wider into space. This will result not in linear growth, but in cubic growth. Yet Malthus was essentially right: exponential growth will outrun cubic growth as easily as it would linear. Calculations show that unchecked population growth, with or without long life, would overrun available resources in about one or two thousand years at most. Unlimited exponential growth remains a fantasy, even in space.

Civilizations

Do other *civilizations* already own the resources of the universe? If so, then they would represent a limit to growth. The facts about evolution and technological limits shed useful light on this question.

Since many Sunlike stellar systems are many hundreds of millions of years older than our solar system, some *civilizations* (if any substantial number exist) should be many hundreds of millions of years ahead of ours. We would expect at least some of these civilizations to do what all known life has done: spread as far as it can. Earth is green not just in the oceans where life began, but on shores, hills, and mountains. Green plants have now spread to stations in orbit; if we prosper, Earth's plants will spread to the stars. Organisms spread as far as they can, then a bit farther. Some fail and die, but the successful survive and spread farther yet. Settlers bound for America sailed and sank, and landed and starved, but some survived to found new nations.

Organisms everywhere will feel the pressures that Malthus described, because they will have evolved to survive and spread; genes and memes both push in the same direction. If extraterrestrial civilizations exist, and if even a small fraction were to behave as all life on Earth does, then they should by now have spread across space.

Like us, they would tend to evolve technologies that approach the limits set by natural law. They would learn how to travel near the speed of light, and competition or sheer curiosity would drive some to do so. Indeed, only highly organized, highly stable societies could restrain competitive pressures well enough to avoid exploding out ward at near the speed of light. By now, after hundreds of millions of years, even widely scattered civilizations would have spread far enough to meet each other, dividing all of space among them.

If these civilizations are indeed everywhere, then they have shown great restraint and hidden themselves well. They would have con trolled the resources of whole galaxies for many millions of years, and faced limits to growth on a cosmic scale. An advanced civilization pushing its ecological limits would, almost by definition, not waste both matter and energy. Yet we see such waste in all directions, as far as we can see spiral galaxies: their spiral arms hold dust clouds made of wasted matter, backlit by wasted starlight.

If such advanced civilizations existed, then our solar system would lie in the realm of one of them. If so, then it would now be their move—we could do nothing to threaten them, and they could study us as they pleased, with or without our cooperation. Sensible people would listen if they firmly stated a demand. But if they exist, they must be hiding themselves—and keeping any local laws secret.

The idea that *humanity* is alone in the visible universe is consistent with what we see in the sky and with what we know about the origin of life. No bashful aliens are needed to explain the facts. Some say that since there are so many stars, there must surely be other civilizations among them. But there are fewer stars in the visible universe than there are molecules in a glass of water. Just as a glass of water need not contain every possible chemical (even downstream from a chemical plant), so other stars need not harbor civilizations.

We know that competing replicators tend to expand toward their ecological limits, and that resources are nonetheless wasted through out the universe. We have received no envoy from the stars, and we apparently lack even a tolerably humane zookeeper. There may well be no one there. If they do not exist, then we need not consider them in our plans. If they do exist, then they will overrule our plans according to their own inscrutable wishes, and there seems no way to prepare for the possibility. Thus for now, and perhaps forever, we can make plans for our future without concern for limits imposed by other civilizations.

Growth within Limits

Whether anyone else is out there or not, we are on our way. Space waits for us, barren rock and sunlight like the barren rock and sun light of Earth's continents a billion years ago, before life crept forth from the sea. Our engineers are evolving memes that will help us create fine spaceships and settlements: we will settle the land of the solar system in comfort. Beyond the rich inner solar system lies the cometary cloud—a vast growth medium that thins away into the reaches of interstellar space, then thickens once more around other star systems, with fresh suns and sterile rock awaiting the touch of life.

Although endless exponential growth remains a fantasy, the spread of life and civilization faces no fixed bound. Expansion will proceed, if we survive, because we are part of a living system and life tends to spread. Pioneers will move outward into worlds without end. Others will remain behind, building settled cultures throughout the oases of space. In any settlement, the time will come when the frontier lies far away, then farther. For the bulk of the future, most people and their descendants will live with limits to growth.

We may like or dislike limits to growth, but their reality is independent of our wishes. Limits exist wherever goals are clearly defined.

But on frontiers where standards keep changing, this idea of limits becomes irrelevant. In art or mathematics the value of work depends on complex standards, subject to dispute and change. One of those standards is novelty, and this can never be exhausted. Where goals change and complexity rules, limits need not bind us. To the creation of symphony and song, paintings and worlds, software, theorems, films, and delights yet unimagined, there seems no end. New technologies will nurture new arts, and new arts will bring new standards.

The world of brute matter offers room for great but limited growth. The world of mind and pattern, though, holds room for endless evolution and change. The possible seems room enough.

Views of Limits

The idea of great advances within firm limits isn't evolved to feel pleasing, but to be accurate. Limits outline possibilities, and some may be ugly or terrifying. We need to prepare for the breakthroughs ahead, yet many futurists studiously pretend that no breakthroughs will occur.

This school of thought is associated with *The Limits to Growth*, published as a report to the Club of Rome. Professor Mihajlo D. Mesarovic later coauthored *Mankind at the Turning Point*, published as the second report to the Club of Rome. Professor Mesarovic develops computer models like the one used in *The Limits to Growth*— each is a set of numbers and equations that purports to describe future changes in the world's population, economy, and environment. In the spring of 1981, he visited MIT to address "The Finite Earth: Worldviews for a Sustainable Future," the same seminar that featured Jeremy Rifkin's *Entropy*. He described a model intended to give a rough description of the next century. When asked whether he or any of his colleagues had allowed for even one future break through comparable to, say, the petroleum industry, aircraft, auto mobiles, electric power, or computers—perhaps self-replicating robotic systems or cheap space transportation?—he answered directly: "No."

Such models of the future are obviously bankrupt. Yet some people seem willing—even eager—to believe that breakthroughs will suddenly cease, that a global technology race that has been gaining momentum for centuries will screech to a halt in the immediate future.

The habit of neglecting or denying the possibility of technological advance is a common problem, Some people believe in snugly fitting limits because they have heard respected people spin plausible-sounding arguments for them. Yet it seems that some people must be responding more to wish than to fact, after this century of accelerating advance. Snug limits would simplify our future, making it easier to understand and more comfortable to think about. A belief in snug limits also relieves a person of certain concerns and responsibilities. After all, if natural forces will halt the technology race in a convenient and automatic fashion, then we needn't try to understand and control it.

Best of all, this escape doesn't *feel* like escapism. To contemplate visions of global decline must give the feeling of facing harsh facts without flinching. Yet such a future would be nothing really new: it would force us toward the familiar miseries of the European past or the Third World present. Genuine courage requires facing reality, facing accelerating change in a world that has no automatic brakes. This poses intellectual, moral, and political challenges of greater substance.

Warnings of bogus limits do double harm. First, they discredit the very idea of limits, blunting an intellectual tool that we will need to understand our future. But worse, such warnings distract attention from our real problems. In the Western world there is a lively political tradition that fosters suspicion of technology. To the extent that it first disciplines its suspicions by testing them against reality and then chooses workable strategies for guiding change, this tradition can contribute mightily to the survival of life and civilization. But people concerned about technology and the future are a limited resource. The world cannot afford to have their efforts squandered in futile campaigns to sweep back the global tide of technology with the narrow broom of Western protest movements. The coming problems demand more subtle strategies.

No one can yet say for certain what problems will prove to be most important, or what strategies will prove best for solving them. Yet we can already see novel problems of great importance, and we can discern strategies with varying degrees of promise. In short, we can see enough about the future to identify goals worth pursuing.

7

HANDLE ON THE FUTURE

There have been several attempts at setting up electronic mail within the federal post office administration, but all have failed, not for technical reasons, but because the post office is resistant to change, as it is locked into a *bureaucratic system* in which patronage and civil-service job security are far more important than efficiency. Given that history, it seems to me more likely that electronic mail will come about through private rather than governmental action. At the present rate of development, electronic mail is probably no more than one or two decades away.

Things change. In the distant pre-technological past, Ecclesiastes' "preacher" could state that "there is nothing new under the sun," because the rate of change didn't allow one to see big differences in a single lifetime. That's no longer true.

By and large we predict the future by seconds or minutes, as in making a cup of coffee, or hours and days, as in buying tickets for a ball game. Foresight, as in knowing that you'll be asked for a ticket at the gate when you get to the stadium, consists largely of projecting into the future the situations and results you've seen in the past. The ability to manipulate this kind of knowledge is intelligence, and a reasonably comprehensive set of experiences to go with that ability makes it into common sense.

Every general futurist gets the future wrong to a significant extent more than a couple of decades ahead. The most common failing, by far, is the simplest one: to assume the future will be like the present. Not surprising; that's how our basic future—predicting mechanism works, and it seems to require abandoning common sense, our store of experience, to predict anything else.

The point is important enough to repeat: things don't change that much or that fast in some ways, but in technology they do. In 1981, Gerard K. O'Neill, the Princeton physicist, wrote a book of futurism. In his introduction he reviewed a collection of futurist writings from the previous century and noted that they tended to overpredict political change, but underpredict technological change. Arthur C. Clarke's classic work of prediction, *Profiles of the Future*, likewise begins with two chapters quoting hilariously myopic predictions by leading scientists regarding the impossibility of the electric light, railroad travel faster than fifteen miles per hour, heavier—than—air flying machines, and so forth.

If we turn to contemporary speculation in order to gain some idea of men's expectations of the technology of their times, we find that in their predictions of the future, or rather their extrapolations of contemporary technological trends as they interpreted them, writers often made shrewd prophecies. Following the inventions of telegraphy and telephony, television could readily be imagined; air travel by heavier—than—air machines (usually powered by steam—engines and therefore boasting handsome funnels) was confidently predicted long before the Wright brothers' first flight. Even the atomic bomb was, it is claimed, forecast by H.G. Wells not very long after the beginnings of *sub-atomic physics*.

It is, however, a truly remarkable fact that on the very brink of an economic-technological revolution unparalleled in history no one foresaw the *universal* motor car and all that it was soon to imply. This failure on the part of informed and perceptive men to grasp the significance of what was going on under their very noses must make us suspicious of all attempts to forecast technological developments even one or two years ahead, much less ten or twenty.

What fooled the prophets of 1900 was not the concept of the motorcar—that was well understood. It was the notion that the burgeoning productivity of the industrial revolution might carry through and provide the power of machines to a wide majority of the people. This was perfectly foreseeable: the numbers allowed for it, examples of various kinds were there. One of the major products of the earlier parts of the industrial revolution was cloth, which was made much cheaper and available to everyone. But it defined common sense that the average person could ever afford a complex, powerful, expensive machine not unlike a locomotive.

To do better, we have to tease apart the things that change and those that don't. The best set of unchanging facts we know is science:

basic physics, chemistry, biology, and the like, and their handmaiden, mathematics. Another thing that doesn't change much is human nature, economics, and politics. One gets the feeling, on reading Aristotle's *Politics* and *Physics*, that he would have been flabbergasted by today's technology, hut right at home with our politics.

The things that change are sometimes random, in which case there's not much for us to say. But change often occurs within bounds or along predictable curves. Nobody can say who's going to have a baby next year, but population on the average tends to increase in a smooth exponential progression. It may or may not rain next July 4, but it will be warmer than January 4 (in North America, anyway). I don't know what specific inventions will he made next year, but the progression toward more complete control of physical phenomena is as old as the human race. Every year for the past century there have been inventions and discoveries that allowed us to see more clearly, manipulate more finely, and understand more deeply the physical world. This is not suddenly going to stop.

When we put that technological trend up against a model, we can get a reasonable prediction for future technology. The model in this case is basic science. Technology will get more and more capable, but it won't be able to do things that are impossible. Physics and chemistry will certainly accumulate more knowledge over the coming years. It's possible that there will be new phenomena discovered that will make some things we now believe impossible, possible; it's not so likely the other way around. The basic theory of what goes on at the human to the planetary scale—*Newtonian physics* with *Maxwellian electromagnetic*—has held up for well over a century, and the theory of what goes on at atomic scales, quantum mechanics, has been around for a lifetime. We need only imagine the relentless improvement of technological capabilities reaching the limits imposed by scientific law as we understand it, and we'll have a decent first cut at what a future technology could look like.

Current Trend

Moore's Law is a rule of thumb regarding computer technology which, in one general formulation, states that the processing power per price of computers will increase by a factor of 1.5 every year. This rule of thumb held true from 1950 through 2000. The factor of a billion improvement over the period is nearly unprecedented in technology. The end of Moore's Law has been predicted many times since the 1960s, but it's always been a false alarm. Once doped-silicon field-effect transistors (what's in your current computer) hit a size and speed limit, nanoelectronics can take over. It's already known how to build a molecular switching element, or rather, several ways are known. Just to prove the point, K. Eric Drexler designed mechanical switching elements that could be used to build a gigahertz computer the size of a bacterium.

In other words, don't expect Moore's Law to end in the next two decades. In fact, if you project the Moore's Law curve, you'll find it as good a guide as any to the expected development curve of nanotechnology. The key point here is that what's increasing so fast is processing power *per dollar*. That means that on the same budget, you can do 50 times as much computation as you could a decade ago, and 2,500 times as much as you could two decades ago.

Among other effects, this explosion of processing power, along with the Internet, has made the computer a tool for science of a kind never seen before. It is, in a sense, a powered imagination. Science as we know it was based on the previous technology revolution in information: the printing press. The spread of knowledge it enabled, together with the precise imagining ability given by the calculus, produced the scientific revolution in the seventeenth and eighteenth centuries. That in turn led to the industrial revolution in the nineteenth and twentieth centuries.

The computer and Internet are the calculus and printing press of our day. Our new scientific revolution is going on even as we speak. The industrial revolution to follow—nanotechnology—is just beginning. If you extend the trend lines for Moore's Law a few decades, part sizes are expected to be molecular and the performance-price ratios become nearly astronomical. If you project the trend line for power-to-weight ratios of engines, which has held steady since 1712 going through several different technologies from steam to jet engines, we expect molecular power plants in the 2030—2050 timeframe.

The result of this will be essentially a reprise of the original industrial revolution, a great flowering of increased productivity and capabilities, and a concomitant decrease in costs. In general, we can expect the costs of high-tech manufactured items to follow a downward track as computers have. We can make such a prediction with relative confidence since we have the long-established trend and the scientific model that shows we have headroom for the trend to continue.

When a cat stalks through close spaces in near darkness, she avoids getting stuck in tight passages with her whiskers. They stick

Fig. 7.1. Cat.

out the sides of her face and allow her to judge the width of openings by feel. Many of the designs and devices you'll learn about in the rest of this hook aren't exactly predictions of future technology. They're cat's whiskers. The trend line for engines that indicates that we should have ones with ignore than a million horsepower per pound by 2050. But we can't know that physical law allows such a thing unless we come up with some kind of design and analyze it. By the time most of the gadgets we talk about here can actually be built, our specific designs will probably look pretty old-fashioned and clunky. But our designs feel out the headroom in physics to let us know that there's nothing to stop trends like Moore's Law.

Our designs are like Babbage's Difference Engine in some ways. Once computers got going, no one built them out of ratchets, shafts, and gears. But we could. We don't build them that way because we have something better. But Babbage could know with confidence, more than 150 years ago, that a computer at least as good as his design could be built, and indeed he was right. Where the whiskers fit, there will sooner or later come the cat.

Shape of Things to Come

Nanotechnology is specifically the technology we predict when the tide of technological progress washes against the shore of atomic physics (the quantum mechanics of electrons, with nuclei considered as unchangeable, primitive particles). Other technologies further in the future may well involve other shoals of knowledge, such as nuclear and subnuclear particle physics, and relativity. But these will come later and are beyond the scope of this book.

So what can we do with just atoms? Turning lead into gold is out—transmutation of elements involves changing the nuclei. Flash Gordon spaceships are out—nothing the size of a panel truck can hold enough energy in chemical form to travel from one planetary surface to another unassisted. Psychic powers? Nope, well outside the bounds of known physical law. Antigravity, as apparently used by Superman and the giant flying saucers in various science fiction movies? Sorry!

On the other hand, nanotechnology could turn sewage into breakfast. Transcontinental flying cars are a distinct possibility. Total, all—senses virtual reality, done by injecting signals into your sensory nerves or brain, is on its way. Towers not just miles high, but hundreds of miles high, are quite feasible. The countertop synthesizer or "matter printer" will revolutionize the way you acquire household objects and food.

Consider this: Your waste stream—garbage and sewage alike—is transformed into enriched soil by decay bacteria. Grass and trees grow on the soil. Cows eat the grass. The same atoms you discarded are rearranged into steak, milk, wood, and apples by natural molecular machinery. We can build molecular machinery to do the same thing. Coal into diamonds? Almost trivial by comparison.

Turning sick bodies into healthy ones or aging bodies into young ones is another transformation involving the rearrangement of atoms. These cases are quite a bit more complex than the previous ones, since we don't yet understand what actually needs to be done on the molecular level. But nanotechnology will first give us the tools to find out, and then the tools to do the job.

Will nanotechnology be able to build anything that's physically possible, as far as the arrangement of atoms is concerned? Probably not. First, for some arrangements that would seem to he physically stable if they did exist, there may be no physically possible way to get there from here.

A much larger class of structures are those that seem theoretically possible, but aren't in the practical world. For example, Ralph Merkle has pointed out that physics has no problem with a cube of diamond a meter on a side (i.e., a flawless three-and-a-half-ton diamond), which is a perfect crystal down to the atomic level. And in fact, nanotechnology could (and probably will) make such a diamond cube. In practice, though, it could not be perfect at the atomic level, because cosmic rays (highly energetic ionizing particles that are sleeting through our bodies all the time) would knock plenty of the atoms out of place.

Yet a larger class of things nanotechnology won't make are the ones requiring the development of new manufacturing techniques that, while quite feasible, don't have any useful application. Similarly, our current technology could make, in theory, many things for which the particular machines and techniques necessary will never be developed. (For example: a single-crystal pure silicon coffee mug, whose inner surface is lined with circuitry etched into the silicon just the way it is on a chip. That's clearly something current technology is capable of in theory, but it would require retooling a $10 billion fabrication plant to actually build.)

And of course, the biggest category of things nanotechnology won't make is the group of things that are possible, feasible, useful, and economical, but which we just aren't smart enough to think of.

Expected Time

It's possible that nanotechnology already exists as the result of a secret government effort like the Manhattan Project. Whether such a project could have succeeded by now depends on a whole raft of variables. It is getting close to the point that it would be irresponsible not to have such a project. However, it is not clear that such a secret military attainment of nanotech capabilities would make all that much difference to the course of commercial development. It would not accelerate development, since the military version would be kept secret. There probably would not be much hindrance, either, since any significant activity of that nature would both arouse suspicions and hurt US economic competitiveness in the relatively near term.

So we can hope that secret early development happened, as a hole card against unexpectedly early development by malicious forces else where in the world. But most likely, we can ignore the possibility without invalidating our analysis of mainstream development.

The road to nanotechnology; as Drexler described it in his *PNAS* paper, and in much greater detail in *Nanosystems*, was as a more or less straightforward extension to biotechnology. We already know the genetic code, that is, we can specify a sequence of DNA that will, tell the ribosomes in a cell to create a protein with a specific sequence of amino acids. It is the sequence that causes a protein to fold up into a specific shape, and the shape that determines its function as a molecular machine.

Determining the shape that a protein will food into is called, appropriately enough, the "protein folding problem." It is one of the holy grails of current molecular biology (part of the subfield known as proteomics). When Drexler's paper appeared in 1981, it was essentially an open problem. At this writing, good techniques are known for special cases, and plenty of progress is being made.

However, as Drexler pointed Out in 1981, there is a big difference between science and engineering. It is a distinction often lost on journalists, so when they need a pronouncement on the likelihood of some technological advance or another, they have a tendency to ask a scientist. This is often a mistake.

Science, historically, tends to lag behind engineering. People had been burning candles for millennia before Michael Faraday's famous lectures. The makers of wine and cheese had been using microbes for centuries before Louis Pasteur identified them. Steam engines had been in use for a century before Sadi Carnot elucidated the principles of thermodynamics. Celluloid and Bakelite were in use for decades before chemists finally understood that polymer molecules were long chains of their monomer constituents.

Both science and engineering involve the discovery of natural law and its application to the real world. The difference is that the scientist is trying to discover the most general law possible, applicable to any situation, while the engineer is trying to discover recipes to build machines that work in exactly one situation: when built as specified.

Imagine it is the 1600s and we are comparing Sir Isaac Newton with a military engineer cf the day. Newton's law of gravity, one of the towering achievements of the human intellect, is applicable to all masses anywhere in the universe. The engineer is interested in a cannonball of exactly ten pounds traveling up to a mile from his gun. He can tell you how much powder to put in the gun and what elevation to give the barrel—because he is tried it. Newton's theory cannot do that, because nobody knows how much energy the exploding powder produces, or how efficiently that energy is transferred to the cannonball, or how much much the bail is slowed by friction with the barrel or the atmosphere on its flight. The engineer does not know these things, either, but he does know how much powder to use to shoot a mile. The chemistry and aerodynamics (supersonic aerodynamics!) necessary to figure all this out from first principles didn't come along for centuries.

Similarly, as Drexler observed, you don't have to solve the protein folding problem in general to do the opposite: design a protein that will fold up a specific way--called de novo protein design. You just use the sequences of folding you understand, and do not use the ones you do not. Sure enough, within a decade, it had been done, and by now it is a substantial field of its own.

The next step is to design proteins that do not just sit there, but do something. A protein that does something is called an enzyme. Progress is being made in this field as well, although we still, cannot call up a protein engineer and ask for a protein that will catalyze a given reaction.

Where does that get us? For an analogy, imagine that you are an explorer who has crashed on a hidden plateau like Sir Arthur Conan Doyle's Lost World. Instead of dinosaurs, however, what you find is a whole civilization of friendly, helpful people with an advanced technology. The catch is that their technology is based entirely on plastic. They have, let us say, nylon, polyethylene, and styrene. They can make any shape they like in these plastics and have a broad, highly capable technology based on them. They can make cloth and clothing. They print books. They have beautiful buildings of outsized Lego blocks. They have windmills, waterwheels, and even large but underpowered steam engines, which they use to drive stately boats slowly across their placid waterways.

But you'd like to get home, and to do that, you need an airplane. You can build many parts of an airplane from their materials, but you need metal for the engine: no engine of plastic has anywhere near the power-to-weight ratio you need. They have no metal—but the ground, you notice, is rich with metal ores, and there's plenty of clay. What you do next is a fun little exercise in imaginary engineering. Basically you do something like have your friends build plastic molds that you can use to build clay molds and crucibles with which you can make metal parts. Then you work your way up to a machine shop in which you can make the precision parts needed for an engine. Depending on the specific ores available, the process could go lots of different ways, all of which would lead in the end to a metal-based metalworking technology.

The plastic technology is, of course, life. Biotechnology has a complete, self-sufficient manufacturing base, which is even self-reproducing. All we need to do to get to full-fledged nanotechnology is to bootstrap it to produce the kind of parts that can move it out of its current water-based, temperature-sensitive strength and power-poor limitations. This approach is called the bottom-up approach to nanotechnology

View from a Height

Another person who foresaw nanotechnology, in broad outline, was Richard Feynman, the second-most famous physicist of the twentieth century. In a now-celebrated talk given in 1959, he explained why "There's Plenty of Room at the Bottom." Much of the talk sounds quaint now, more than four decades later: Feynman foresaw things like the advantages of ultraminiturization in electronics for computers that we take for granted today. You have to remember that at the time computers with discrete transistors were the very latest thing, and most still used vacuum tubes.

Feynman had a different notion about how to proceed to a broadly capable technology at the molecular scale: start with the broadly capable technology we already have at the everyday scale, and make it smaller:

> Now comes the interesting question: How do we make such a tiny mechanism? . . .[I]n the atomic energy plants they have materials and machines that they can't handle directly because they have become radioactive. To unscrew nuts and put on bolts and so on, they have a set of master and slave hands, so that by operating a set of levers here, you control the "hands" there, and can turn them this way and that so you can handle things quite nicely.
>
> Well, you get the principle from there on. It is rather a difficult program, but it is a possibility.

This is called (*surprise*!) the top—down approach. Like life, it is autogenous. An autogenous technology does not necessarily mean a specific self-reproducing machine. It more generally means a set of manufacturing equipment that includes as a subset of its output everything that is necessary to make more of each kind of the equipment. No industrial base could survive without this property: some unmanufacturable machine would break, and things would grind to a halt.

Let's consider what the pure Feynman—style path to nanotechnology might look like. Essentially we would build an autogenous system in whatever macroscale technology we could, and then use it to rebuild itself iteratively, but on a smaller scale each time.

This is not as simple as it sounds, or it would have been done long ago. It's tough enough to have a milling machine produce objects as precise as its own parts, for example much less improve them by a factor of four. Then one must have a design that is scale—invariant, or a way to change designs as scaling laws invalidate techniques. One of the advantages of this capability—first, size—second approach is that when scale—affected techniques do give out, you have a broadly capable system you can use to experiment with alternative techniques.

The system must be able to do handling and assembly as well as fabrication, since those operations cannot be done by a human operator as in a macroscopic machine shop. It must be able to "see" what it is doing, for the same reason. Feynman proposed using what are now called "waldoes" (after the Heinlein story).' Nowadays, robotics and teleoperation is somewhat more advanced, so we might imagine something more sophisticated—or not; whatever works.

The trick, of course, is to design robots as simply as possible, to reduce the complexity of the system. While we're at it, we need to pick techniques that can scale down as far as possible. An example is to build robots that operate by feel instead of vision, even though that may be slower and less efficient at the macroscale: at the nanoscale, vision won't work but touch does. Electrostatic motors work at the nanoscale but electromagnetic ones don't.

One difficulty for a machine, or a shop full of machines, that tries to build a copy of itself, is that all the design trade-offs that help you make precision parts tend to force you to build parts smaller than the manufacturing machine. One way around this involves a two-phase system which has big fabricators that make small parts, and assembly robots that put the parts together into fabricators and robots. (In a sense, that is how a standard machine shop works, with people as the robots, except the way people make other people isn't by putting parts together.) Such an architecture can be fairly scale-independent.

So how many steps would it take to get to molecular scale? The first thing to realize is that you have reached molecular scale not when your parts are the size of atoms, but when your tolerances are. A typical fine machining accuracy for most of the twentieth century was a ten—thousandth of an inch (2.5 microns); there are specialty shops that will do 1 micron. Atomic precision is about an angstrom (one—tenth of a nanometer), so you need to improve by a factor of ten thousand. If we can quarter the tolerance at each stage, as Feynman suggests, we are about seven steps away How long would it take for each step, in a well-funded project? Maybe a year, maybe two.

Given current technology, though, you can skip way ahead. The electron beam machines currently used for nanolithography can cut shapes with a precision of about 2.5 nanometers. That's a thousand times better than the machine shop standard. With an e-beam machine, as they are called, you're only about a factor of twenty-five away from atomic precision: two factor-of-five steps or three factor-of-three steps. The problem is that there is not a full machine technology at that scale. So it is a lot harder to take the shapes you have cut and put them together into a machine. Starting with a machine shop or nanolithography are the extremes of one set of choices—you can jump in at any scale in between. The smaller you start, the fewer steps to atomic scale, but the more limited your manipulation capabilities are initially.

Ultimately, the question of when devolves into the questions of how much effort is put into which pathways. If the billion dollars a year being spent by the National Nanotechnology Initiative were focused on molecular manufacturing in a well-directed way, we could almost certainly he at Stage III within ten years, and at Stage V in another ten. An Apollo Project—scale effort might cut that to five and five. But none of that is happening. The NNI's billion is being spread over a wide variety of science and technology, and the only thing it all has in common is that something about each piece can be measured in nanometers. That's just the way political funding works.

That's not to say the money is being wasted, by any means. It's going to good and valuable research in areas that have shown promise. The US government wastes billions each year, but this isn't where. It's just that a well-conceived, sharply focused effort might well succeed faster with less money.

Destination

There was an amazing flowering of science in France at the dawn of the nineteenth century. D.S.L. Cardwell writes, "During the years 1790—1825 France had more scientists and technologists of first rank than any other nation ever had over a comparable period of time." We can mention Sadi Cannot, Antoine Lavoisier, Pierre-Simon Laplace, Joseph and Jacques-Etienne Montgolfier, Pierre Dulong, Alexis Petit,

Jean-Baptiste Biot, Augustin Fresnel, Joseph Gay-Lussac, Andre Ampere, Felix Savart, Joseph Fourier, Gaspard-Gustave Coriolis, Augustin Cauchy, and Jean Lamarck—and these are just the ones whose names are attached to scientific laws and inventions that have survived to the present.

What's more, the advancement of science and technology was a well-funded national policy. Carnot was a graduate of the Ecole Polytechnique, an institution that had no parallel in England or anywhere else at the time. England by contrast, far from supporting its scientists, allowed top-caliber people to be bounded for religious or other reasons. While the leading scientists of France were given seats at the Ecole or other institutions and expected to teach the next generation of French scientists, Joseph Priestley, the discoverer of oxygen, was run out of England for his religious beliefs, and lived out his life in America.'

Besides the names, the textbooks and technical literature of France were notably superior. Charles Babbage, forefather of the computer and holder of the Lucasian chair at Cambridge (the professorship Newton had held), writes of searching for French calculus books and starting a society for the adoption of the Leibnizian notation over the cumbersome Newtonian one.

It was not only an academic leadership. France was the acknowledged leader in most fields of actual technology. From the Jacquard loom, an early example of automated control, to the invention of the balloon and parachute, the French were ahead. Their roads, bridges, and cathedrals were better. They even built more advanced ships than the British at the time. French policy included public recognition and prizes for scientific and technical discoveries, and public funds were available for the development of new inventions.

So if you are a technological forecaster, what do you think happened next? What actually happened, of course, is that the industrial revolution occurred in Britain, not France. By 1850, Britain had rail roads; Britain had steamships; Britain had the leading engine, machine tool, and textile industries in the world.

If you plot the amount of warfare going on all over the world from 1500 to 2000, you'll notice an odd fact. It holds a fairly constant average level throughout the period, with the exception of the nineteenth century. The level of warfare was consistently and significantly lower in the 1800s. This was the Pox Britannica: Britain held a global empire with a technology so much more advanced that it was not much challenged. It is not that the British were particularly peaceful: they began the century in the throes of the Napoleonic Wars, and ended it in the Boer war. Nor were they particularly more warlike than average, either. But they had the edge, so the world at large enjoyed a century of relative peace.

It is of more than academic interest to understand how this happened. The industrial revolution of nanotechnology has the potential to wreak a greater change on society than the one of steam. And it is clear that, as the players enter the stage for our rerun of this little historical drama, the United States is playing the role of France.

Historians have debated the issue endlessly and offer reams of theories. Those that ascribe some advantage to England per se, however, we can dismiss. Fast—forward to 1900 and we find British technological dominance in decline, and America on her way up. More likely is the analysis that notes the centralized structure of French science and technology. Paris was a brilliant center of learning, but it drew the best minds from the rest of the country. In some sense, the French system developed an ivory tower and insulated it from the practical problems on the ground. Thus, great strides were made in theoretical science; but practical techniques, which did not win the acclaim of the learned, received little attention. In England, almost perversely because of the official indifference to science, the best minds remained in contact with the real world.

America today has a centralized science funding system and the effects can be seen in the political distribution of nanotechnology funding. One such effect well understood, and completely expected—is that most of the funding labeled for nanotechnology is channeled to people who are well—established scientists doing more or less the same things they always did. Lots of good and useful things will be done in these labs, and they will in fact ultimately lead to the kind of nanotechnology we are interested in here. But they won't do it as fast as the same amount of effort focused directly on nanomachine systems would.

In 1995 the number of doctorates in engineering earned in Asian universities exceeded those from US universities—and more than half of the engineering degrees in American universities were already being earned by students from Asia. Engineering doctorates from Asian universities continue to rise at ignore than 10% per year, but American engineering doctorates are actually declining. It is all too easy to imagine a scenario in which India, China, Hong Kong, Singapore, Tiawan, South Korea, and Japan are competing to build nanomachines while the US funding agencies are still trying to soft-pedal them for fear of a genetically modified foods-style backlash. Asian populations have more to gain, less to lose, and in many cases, less say in the matter, than Americans.

Americans have been in this kind of science slump before, and were startled awake by the sight of *Sputnik* serenely cruising over their heads. If we full behind on nanotechnology, however, there is very little chance we could catch up. Imagine that the wake—up call is break through into Stage III nanotech—that someone has a mechanical version of a ribosome that can build copies of itself and other nanomachines out of molecular building blocks. Whoever has such a machine holds an enormous advantage over those who, like today nanotech researchers, are still building every nanomechanism from scratch. Stage III systems could make nanoscale measurement equipment, test fixtures, and experimental prototypes of new machines—the technology base from which to build Stage IV systems.

The road to nanotechnology is like a five-mile race. You have to swim the first mile. Then you can run the second. At that point you find a bicycle. At the end of the third mile is a car, and at the end of the fourth is an airplane. The United States has more people in the water at the moment, but they're swimming in a broad group at random directions, mostly toward short-term goals, and not toward the next stage. If someone gets a significant lead, he'll he hard to catch.

Failure of the Imagination

When a distinguished but elderly scientist states that something is possible, he is almost certainly right. When he states that something is impossible, he is very probably wrong.

Scientists, in general, make lousy futurists. The job of a scientist is to establish facts by careful experimentation, documentation, and repeated testing until it is as verified as human inquiry can make it. If futurists waited for the kind of verification scientists require, they'd be historians.

In fact, a historian, particularly a historian of technology, makes a much better futurist than a scientist. Beyond the immediate horizon of what's in the lab today, the future is shaped by the same forces that shaped the past, and the past is one of our best guides for prediction.

Future technology is shaped by what people want. It's also shaped, of course, by what is possible. But a lot more is possible than people, particularly scientists, realize. The reason is that there are usually many different ways of getting a given result. A scientist can tell you, quite correctly, that a specific approach using a specific phenomenon won't work. But he can't know whether some other way, outside his narrow field of expertise, would reach the same goal.

An example very germane to nanotechnology is taking pictures of atoms. You can't use an ordinary microscope to take pictures of atoms because atoms are much smaller than the nature of light waves will allow you to focus. If you asked a microscopist, say, a century ago, whether it would ever be possible to take pictures of atoms, he would have said no, and been right by every rule of optics.

X-rays have the right wavelength but they have a couple of problems: there' nothing we can use to focus them and they don't bounce nicely off atoms but either go right through or shove them around. By mid-century people were using x-rays to discover molecular structures, by an ingenious method involving crystallizing millions of molecules into a regular array and interpreting the scatter-shot diffraction patterns x-rays made off the crystals. But they would have told you, rightly, that nice photos of individual molecules, showing the rounded curves of the electronic surfaces, simply couldn't be taken.

Then in 1986, Gerd Binnig and Heinrich Rohrer won the Nobel Prize for doing just that. But they didn't actually look at atoms; they felt them. The scanning tunneling microscope (STM), which they invented, uses a computer to control a needlepoint, atomically sharp, that "feels" the atoms by a very short—range electric interaction. The computer then puts the results together into a picture.

The STM came out of left field; nobody expected it. Once the principle was understood, a huge variety of similar scanning probes was invented, and pictures of atoms are commonplace today. Scanning probes are quite simple to build—they can be, and have been, built by high school students as science fair projects. But the point of interest here is that they use completely different physical principles than anyone would have guessed as the way to image atoms. A scientist trying to predict whether such a thing was possible would have gotten it wrong, ironically, by knowing too much about the way previous imaging technology worked.

Similarly, the laboratory scientist of today can tell you, quite rightly, that current fabrication technology can't make molecular machine parts out of diamond. But the historian of technology can point out three—hundred—year trends that point very strongly in that direction.

The futurist can add some careful analyses and computer simulations and make more specific predictions. It is, after all, our profession.

In 1869, when Jules Verne wrote *Twenty Thousand Leagues under the Sea*, seagoing submarines were science fiction. In 1900, flying machines were science fiction. In 1950, space travel was science fiction. In 1963, a Robert Heinlein character "took his phone out of his pocket" while walking in a park to call a friend. Today, an account of the future involving molecular manufacturing and nanomachines could be considered science fiction—but an account of the future without them is fantasy.

8

SELF-REPLICATION

Go to a park, a forest, a farm, anywhere you can see the natural world. Look about you. Virtually everything you see—trees, bushes, grass, cows, mice, insects, birds—is a self-reproducing machine. Now back to your house, apartment, cityscape. Virtually nothing is. The works of man are distinguishable from the rest of the world by being dead: they do not grow; they do not maintain or repair themselves; they must be built.

Nanotechnology can certainly create artifacts that would give houses, cars, and cities some of the better attributes of natural life, such as the ability to maintain and repair themselves, without raising even the possibility of the ludicrous scenarios of cheap horror fiction. Unfortunately, there has been enormous misrepresentation not only from Hollywood type factories but from "serious" commentators who should have done their homework but didn't. So let us see if we can dissipate some of the smoke and shed a little light on the subject.

Some quarter of a million years ago (or more), humans mastered our first self-reproducing technology: fire. Since then, most people have needed to know enough about fire to control it. You needed to know how to start one, maintain it at an even rate for cooking and heating, and not let it spread to engulf clothing and buildings. Traditional fire skills are still taught by the likes of the Boy Scouts. Fire requires three things: fuel, air, and heat. Provide all three to start one; deny any one to put it out. By means of this relatively simple recipe, people have lived in wooden houses and burned wood in fire places for millennia. The vast majority of humans are capable of doing this with perfect safety.

Any self-reproducing technology will have a similar high-level logic. It will require raw materials, energy, and a controlling pattern. All of these will be necessary to start the process, and the removal of any one will be sufficient to stop it.

Most of us still use fire; but we don't even need to know much about it anymore. There is a fire in your furnace, one in your oven, one in your gas grill, and hundreds of fires are lit every second in the cylinders of your car engine. Machinery starts and stops these fires in carefully designed, completely controlled conditions, and in many cases you don't even know they are there.

Our second self-reproducing technology was *agriculture*, coming about ten thousand years ago. With this technology we reshaped the face of the Earth. Farmers labour mightily to provide the self-reproducing plants with the right environment. They do not pick the plants that reproduce most easily, but those with the most useful produce, which are often much harder to grow than the weeds.

Note in fact that people have used *self-reproducing agents* for centuries in applications where no one knew that actual self-reproducing agents were present: *yeast* in wine and bread making, and *bacteria* in

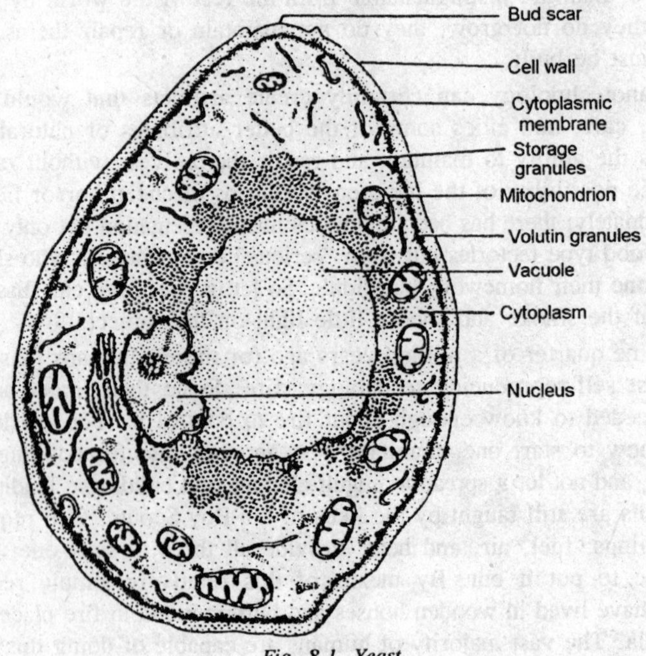

Fig. 8.1. Yeast.

making cheese. Plenty of people have fireplaces; make bread, wine, and/or cheese; grow plants; or raise animals for fun today. Reproduction is a wonderful process but it is not difficult to control, nor dangerous in and of itself. It is not even, all that mysterious.

Automated Production Machinery

By the middle of the twentieth century, scientists were beginning to realize that the functions of life were completely mechanistic and that this applied to reproduction as well. One of the great minds of the period, John von Neumann, whose name is strongly associated with the development of the computer, worked out a rigorous structure for self-reproducing *automata*, as he called them.

The basic idea is that you have *automated production machinery*; which can take raw material and produce more machinery. The machinery has instructions, which in the 1950s was typically a punched paper tape, that guide it in the making of specific parts. A self-reproducing machine would be a kind of machine shop that would be able to produce all the parts of all the machines. It would include a sort of automated assembly line, a controller that read the paper tape to direct all the machines, a (long) paper tape that directed the machines to build another shop just like the first, and a gadget to copy the tape.

Remarkably, not too long after von Neumann worked this out, the biologists who were studying the inner workings of cells figured out that at a high-level description, the way cells reproduced was just like the von Neumann scheme. DNA was the tape, and there were specific gadgets to build parts and to copy DNA.

On the Factory Floor

At a lower level of description, of course, what goes on in a *cell*, does not resemble an assembly line, automated or not. In particular, the parts that a cell produces, protein molecules, are not picked off the production machine and carefully conveyed to meet the next part in the process. Instead they float loose to bump randomly into other parts until they meet one they click together with, in a process called *self-assembly*.

In a nanomachine, as in a human-sized factory, parts are never set loose to drift around. They are moved by robotic arms, shuttles, or conveyor belts from spot to predetermined spot and are assembled by mechanical force. This has two big advantages: things move much faster, going directly to the spot where they are needed rather than

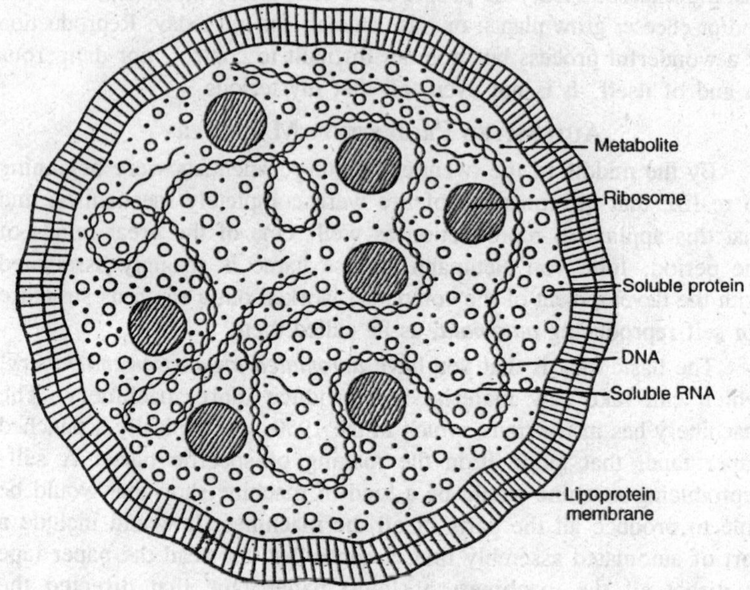

Fig. 8.2. Prokaryotic cell.

having to do a walkabout; and parts can be designed for their function alone, instead of having to be designed for their function and also to pop together by themselves. Designing for self-assembly is a major constraint: not only must the parts that are supposed to fit together do so, but parts that are not intended to go together had better not fit. In a regular factory; if tab A would fit into slot C, where it does not belong, as well as slot B, where it does, it not a problem; you simply don't put it into slot C.

The *nanomachine* is a lot faster and simpler in some ways than the machinery of the cell. On the other hand, it has a big disadvantage: it can't evolve. *Cells* work the way they do in part because it is easy to make small changes that leave the whole thing working. Suppose you want to add a part to an existing machine. All you need to do is start producing it. If you get the shape right, to match the existing machine, the part will bump into the machine, latch on, and start working. In an assembly line context, you need to design not only the new part, but the conveyance equipment, the assembly machinery, the process flow, and the control mechanism or software.

Now remember that in evolution, 999 changes out of 1,000 are for the Worse, and die out. If you try to build a new part and get it wrong, it simply does not fit and nothing much happens. But in the factory model, to have any effect at all, you have to do a whole bunch of changes at the same time and get all of them right the first time.

Imagine an old-fashioned, *open-air bazaar* or *marketplace* instead of having fixed stalls, all the vendors have pushcarts and are constantly wandering around, buying and selling with the other vendors they bump into. Suppose there is an omelette maker. She bumps into sellers of charcoal, eggs, butter, and paper plates (all bought one at a time), at random. When she gets all the necessities, she makes an omelette and is ready to sell it to the next omelette customer she meets.

Now if you wanted to change things a bit, you could throw a few cheese dealers into the bazaar. After a while, you might reasonably expect some cheese omelettes to start showing up, as the omelette maker bumped into the new vendors at the right stage in the process.

Now imagine a *factory* instead. It is a building where people cannot wander around. Trucks with the various ingredients come to specifically designated loading docks, one for eggs, one for butter, and so on. Inside, there is a specially equipped room for each operation, for example, one where eggs are broken, one where pans are buttered, one where half-cooked omelettes are flipped. In each room is an expert who does only that one operation. He gets exactly the inputs he needs out of holes in the wall, without looking at them because they are guaranteed to he exactly the same every time. He does his operation without looking, as well, and passes the result out a hole in another wall.

Now to change this setup to make cheese omelettes, you have got a major problem. To begin with, you have to build a whole new loading dock for cheese. Then you have to build a new room to stir the cheese in between the egg cracking and the cooking. But there's no room! The existing rooms are right next to each other. So you have to tear down half the factory and rebuild it with a new room and dock and so forth in the middle.

A cheese vendor might make a wrong turn and end up in your bazaar by accident, and then keep coming back because it is profitable. Whole, working, factory redesigns and reconstructions just don't happen by accident. Coordinated, integrated *nanomachine systems* with deterministic mechanical materials transfer can be much faster and more efficient than cells at doing some one specific pre-designed thing. They pay for this by giving up the complex, adaptive capabilities necessary for evolution. They can be adaptable if they are predesigned to handle more than one situation or process, but for every extra capability they would become less efficient. They simply cannot evolve by themselves. This is one point the fiction writers either fail to understand or deliberately ignore. They take some properties of living things, like their adaptability; and some properties of mechanical systems, like process speed, and conjure up imaginary monsters that have both. But in the real world you have to give up one to get the other: it's a trade-off.

Reproduction versus Replication

Researchers in the field, following the lead of Moshe Sipper at the Swiss Federal Institute, refer to systems that make copies of themselves as "*replicating*" rather than "*reproducing*." The difference is that reproduction involves making, on purpose, imperfect copies for the purposes of evolution, whereas replication is the simpler mechanical process of making an exact copy. As we have seen, living cells pay a lot for the capabilities they need to evolve.

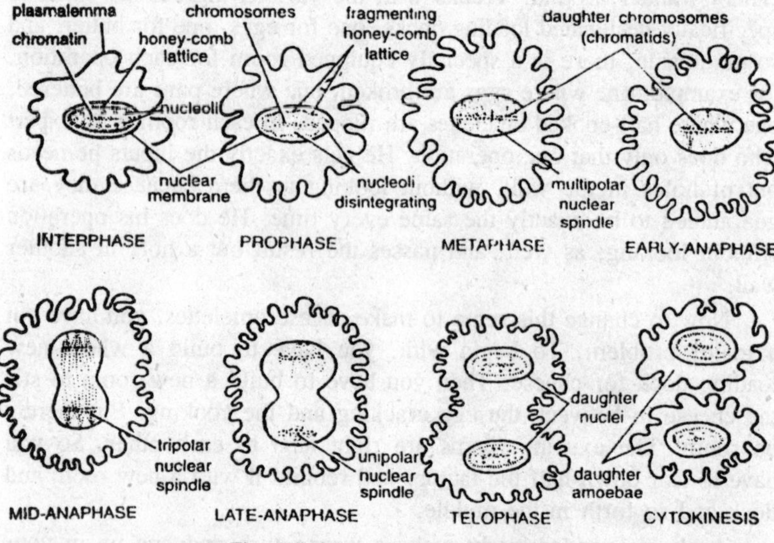

Fig. 8.3. Binary fission in Amoeba.

Another thing that seems fairly necessary for evolution, but not *simple replication*, is sex. In sex, the genetic information describing a given organism comes from two parents, four grandparents, eight great-grandparents, and so forth. Not too many generations back and you are getting information from most of the population of your species. That means that any beneficial mutation from any member has a chance to reach you. With nonsexual reproduction, the mutant would have

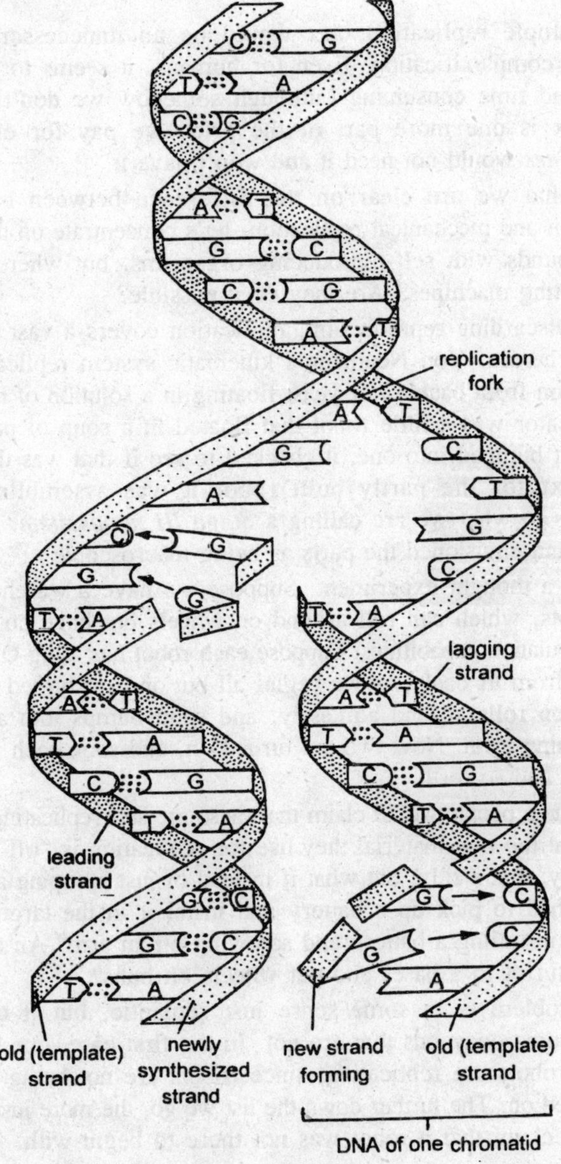

Fig. 8.4. DNA replication.

had to have happened to the single individual that was your ancestor. (Note that even bacteria trade DNA and are "sexual" for purposes of this analysis.)

For simple replication, *sex* would be an unnecessary, time-consuming complexification. Even for humans, it seems to he fairly complex and time consuming (although somehow we don't seem to mind). Sex is one more part of the price we pay for evolution. *Nanomachines* would not need it and won't have it.

Now that we are clear on the difference between biological reproduction and mechanical replication, let's concentrate on the latter. Nature abounds with self-reproducing organisms, but where are the self-replicating machines? Are they even possible?

Even discarding reproduction, replication covers a vast range of different schemes. Von Neumann's kinematic system replicator took its inspiration from bacteria or yeast floating in a solution of nutrients. Each replicator was a little robot that floated in a soup of parts, and each time it bumped into one, it checked to see if that was the one it needed next for the partly built robot it was assembling. This corresponds to what we are calling a *Stage III nanosystem*, although von Neumann envisioned the parts as being macroscopic.

Here's a thought experiment: Suppose we have a warehouse hill of toy robots, which can roll around on wheels but have no arms or other manipulation capabilities. Suppose each robot has a big ON button protruding from its back. As we begin, all but one are turned off. The one that's on rolls around aimlessly, and soon bumps into another's button, turning it on. Now two are turned on, and soon each will turn another on, and so forth.

Now some people would claim that these are self-replicating robots. It is just that the raw material they use for replication is "off" robots. OK, you say, that's silly; but what if instead of just bumping a switch, the robots had to pick up a battery and instal it in the target robot? How about installing a battery and screwing on an arm? An arm to a turret, the turret to a base, and put wheels on hubs?

The problem is in some sense just semantic, but it can help illuminate some questions that are not. In the first case, one hesitates to say the robots are replicating, since robots are not being created, merely turned on. The further down the list we go, the more justification we have to claim that a robot was not there to begin with.

One thing to note about the first, on-switch, example is that although the robots are not replicating, some replicating process is nevertheless present. Although it requires complex inputs and makes minimal changes, robotic activity is replicating. Another thing is that the process is glaringly obviously possible.

As we move down the list to the point where the robots are putting together parts like gears, bearings, and shafts, almost everybody would agree that a new robot is being created. This level of self-replication is also definitely, if not as obviously, possible, because it's been done in robot factories. It's more efficient, and thus more common, to have the parts put together by machinery much more specialized than the robot being built, but it has been done both ways.

Let's go back to the second example, where you have to screw on an arm and install a battery. Imagine that to do both, a robot has to use a battery-shaped gripper, install the battery, change grippers, change positions, install a screwdriver on its arm, screw on the new robot's arm, and then move back and change tools for the next incoming robot. Clearly, two robots working together, not needing to change positions or tools, could do the work more than twice as fast.

You don't need to read any further in economics than Adam Smith's original *Wealth of Nations* to hit the idea that specialization is a major source of efficiency in manufacturing. This is as true today as it was in 1776. In some sense it is more true, since Smith was talking about specialization of craftsmen and tools, and today we can make completely different machines specialized for wildly different operations.

Consider an ordinary printer is attached to your computer. It can produce a page of simple black and white text in a minute (less if it's a laser printer). You could produce a page like that with pens, rulers, and a lot of practice in what, an hour? The printer is much simpler than you are. You can do a vast multitude of things. The printer can do only one thing, but it does it much faster and more efficiently than you can.

The bottom line: no manufacturing system that can avoid it will use general-purpose machines where specialized ones will do. It would have been possible, using roughly the same level of technology as the printer, to have a robot arm that would write out your pages with a pen, and play chess with you, and dust your monitor as well. It would still be orders of magnitude slower than the printer.

What about replication, though? Specialized machines cannot replicate themselves. Not on an individual basis, anyway. But taken as a whole, they can. Every single machine in any factory anywhere was made, in parts, by other machines in factories somewhere. The system as a whole can make any part of itself. The system as a whole can extend or repair itself even though it does not contain a single machine that can replicate on an individual basis. Such systems are called autogenous.

The more diverse the system is—the more completely specialized—the more efficient it is. *Nanosystems* will be able to pack mil lions of gears, levers, pulleys, and so forth into the volume of a human cell—plenty of room for specialization in any system big enough to see. This means *autogenous systems* of coordinated machines, not single replicating ones. Single replicating systems, while technically possible, would be like using robot arms holding pens instead of printers.

Assemblers

Drexler, in *Engines of Creation*, introduced the idea of molecular manufacturing by way of a device called an *assembler*, which was a microscopic, self-replicating, free-floating machine that could use its molecular building capabilities to build products directly after it had built enough copies of itself to make the work go fast.

Assemblers are a conceptual blend of von Neumann's kinematic replicators and existing unicellular life like bacteria and yeast. We do, and have for centuries, make products like cheese and wine by growing hordes of microorganisms in vats. Thus as an explanatory device, the assembler is straight extension and Drexler did not need to expound on the morass of detail and architecture necessary for a more efficient system.

Besides exposition the assembler has one more important role: as a bootstrap. While not as efficient as a printer at printing, a robot arm can build a printer (from the appropriate parts), while a printer cannot build either a robot arm or even another printer. So if you are trying to get started, you'd better build that robot arm first.

Once you get going, you'll have the equivalent of printers, specialized machines that make parts, but you'll still need the robot arms to move the parts around and put them together. Unlike the arms in an assembler, which are precise enough to make sure chemical reactions happen in the right place (i.e., better than an tenth of a nanometer precision), the arms in a parts assembly robot need only be precise enough to put parts together. Parts can be made with bevelled ends, guide flanges, and so forth so that a much looser precision, maybe even a nanometer, will still get everything together exactly.

The reason to build less precise robots is that they can move faster and have a longer reach. You can try this for yourself. Pick up a pen and try writing at arm's length without touching the heel of your hand to the paper, as if you were writing on a blackboard. You'll be a lot sloppier than with your hand touching. However, you can easily and quickly move the pen a few feet, whereas in normal writing position you can move it only a few inches.

Molecular deposition—controlling the chemical reactions to build things to atomic precision, also called mechanosynthesis—is like needlepoint; parts assembly is like stacking firewood. You could build a stack of firewood by gluing together sawdust one grain at a time, but it'd be awful darn slow. That's another point the fiction writers get wrong. A simple assembler that produced something with a single arm doing serial molecular deposition would be considerably slower than a complex factory-like system.

Mechanosynthesis

The *molecular machines* in your cells are atomically precise—two copies of the same protein are atom-for-atom the same—because they are built by a series of controlled chemical reactions. Reactions in the cell don't happen at random, as they do in a test tube or fire, but only where the resulting molecular structure is exactly the needed next step in the construction process.

Similarly, in a *nanomachine*, the process of building a part will consist of a sequence of chemical reactions that produces an atomically precise product. Robot arms, or factory-like machines on conveyors, will put molecules together in such a way that only the desired reaction can happen. The result will be a machine part that is atom- for-atom the same as it's designed to be.

Self-assisted Assembly

The properties of matter at the molecular scale are different from the those we are used to. Part of this is simply size. Compare the flexibility of nylon or polyester fibers with hand-sized solid chunks of the same stuff, which are hard as wood. Even *diamond* at the molecular scale is flexible and springy. *Steel* is like cheese; gold is like dough. *Polyethylene* is like a plate of spaghetti. Liquids are like a load of gravel in a wheelbarrow going down a stairway. *Gases* are like a hailstorm.

Other properties depend on the atomic nature of matter. You can't add half an atom to a part; it's either there or it isn't. Molecular parts are thus bumpy, and it can be difficult or impossible to get shapes like smooth curves.

Finally, atoms are sticky. Van der Waals attraction pulls them together even when there is no chemical bond. It's this force that causes proteins in cells to click together when their shapes match (along with some more complex interactions with water that have a similar effect).

The Van der Waals force is the bone of micromachinery. Together with the fact that they don't have completely smooth surfaces, the result is called *stiction*. It's as if all the parts were made of strong magnets coated with sandpaper. They don't slide very easily. Van der Waals force is even felt in the *macroscale*: it's what holds geckos' feet to the wall. The reason that all objects are not sticky at human scale is that surfaces of macroscopic objects never match at the atomic scale. (The *gecko's* foot has a remarkable form fitting structure that allows it to match the surface it's touching.)

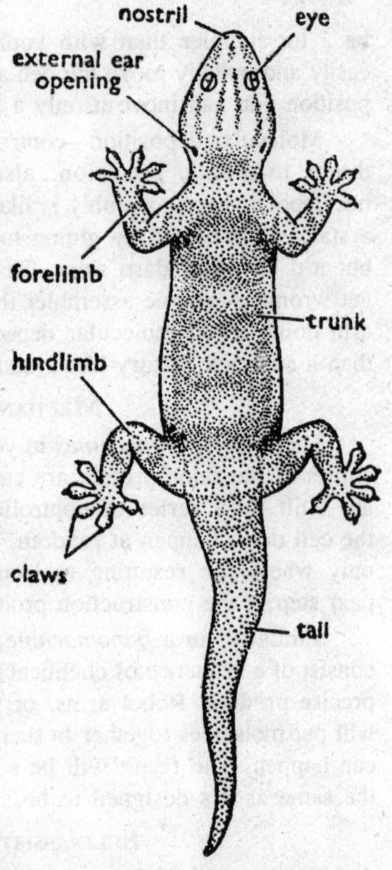

Fig. 8.5. Wall lizard.

Surfaces that are smooth to *atomic precision* will slide, though, despite clinging to each other. So we can make use of the attraction in novel ways. For example, a peg that just fits a hole will be sucked in and will stay put if there's more than a square nanometer or so of surface touching. Thus our parts-assembly robot arm does not have to be very precise at all. Imagine putting a key into a keyhole in the dark. Your hand wanders around, poking at random. At last the key hits the hole and slides in. Your hand does not have to be as precise as the key hole fit; it just has to be precise enough not to put the key in the wrong keyhole. If you had built the keyhole with a funnel-shaped flange, the whole thing would be fairly easy.

The Van der Waals force with properly shaped parts can act like the flange. Indeed it can do more: if you have a screw and a perfectly matching threaded hole, the screw will screw itself in without your having to twist it. Just get the end to the right place. The system of design and construction that uses these properties is called *self assisted assembly*.

Self-assisted assembly does impose some constraints on parts design but not nearly as much as full-fledged self-assembly. For example, you can still use standard designs for fasteners and other interfaces. The arm has enough precision to keep from putting the wrong one in the wrong place, or at the wrong time. In a full self- assembly regime, as in cells, any two flat surfaces will stick together! With self-assisted assembly, we can have all the flat surfaces we like, no problem.

So in a typical *nanomanufacturing system*, you'd expect to find a lot of specialized parts-making devices, producing such things as beams, casings, fasteners, diamond cable, shafts, gears, bearings, motors, wire, and computers. Then there'd be some more general— purpose devices that could make custom parts that did not get called for as often. Similarly, there'd be fast assembly lines to put together parts into commonly used subunits, such as digitally controlled positioners, and more general-purpose robots to put together less commonly called-for assemblies. You'd expect this kind of hybrid architecture not only at the molecular level but all the way up to the size of the output products.

Home Synthesizer

Lets imagine the *home synthesizer* of, say, 2050. It sits on your counter and looks a lot like a microwave. It is connected to your home network, so you program it from your computer, PDA, cell phone, whatever; or just talk to it. You select or create a design with whatever software you prefer; probably most people would use an interface not unlike a Web-based catalog. We'd expect many more options for customization, though, such as ways for clothing to be sized automatically for an exact fit.

The *synthesizer* might be connected to the water and gas pipes and have an "*odd elements*" cartridge that you replaced from time to time. Then you would just tell the computer what you want and wait for the synthesizer to chime and unlock its door. The time taken would be seconds for small, simple items to hours for large, complex ones that involved a lot of nonstandard constructions. By the way, a synthesizer that used natural gas (and air) as raw materials would not use power and water; it would produce them as by-products, pumping them back into the distribution network for use elsewhere. (The exception would be drinks, where more water would be used than produced.)

The *synthesizer* would be able to produce apple baseballs, cups of coffee, diamonds, eggs (for eating, not hatching), folding furniture, gadgets and gizmos galore, headphones, ice cream, jackets, knives (with diamond, not metal, blades), lights, money (paper), nanosuits, office supplies, PDAs, queens (and other chess pieces), robots, synthesizers, tennis racquets (that folded to pocket size when not in use), Utility Fog, vehicles (some assembly required), watches, xylophones (saw that one coming, did you?), yarn, and zircons.

Well, actually, not zircons, since they are *zirconium silicate* and there is virtually no other household use for zirconium, so it would be unlikely to be available in a consumer machine. In fact, it's easy to imagine that economy models might not even handle metals, except for trace amounts in food molecules. Almost certainly a home synthesizer would not be capable of making living creatures, or even eggs that could develop into them.

The reason is that the home synthesizer could not come close to creating any possible arrangement of atoms. It would cover a (tiny) subset sufficient for a large range of consumer products. Another subset with a different background chemistry, namely, fatty acids, sugars, starches, fiber, and proteins, could be used to make foodstuffs. The cells of your synthesized steak might be stuffed with a selection of proteins known to be safe and taste good, but not come close to the full set of molecular machines necessary to function.

Today's paper money would be easily synthesized unless the machine were specifically built not to do it. This is a problem we have seen already, as people photocopied money and printed it from their computers. The official scrip has changed form, albeit slowly, to keep ahead of the counterfeiters, and could easily continue to do so. One simple method would be to incorporate some otherwise-useless element in the official stuff and prohibit home synthesizers from having it. Gold is probably too useful; zirconium may be?

Sorcerer's Apprentice

Since the synthesizer can make another synthesizer, is not it a self-replicating machine? Yes and no—the term is ambiguous. It makes copies of itself, but it does not make copies by itself. When you make a new synthesizer, some person would still have to take it out, set it up, connect it to gas and water, plug it in, and program it before a third- generation one could appear. Thus while the replicator is autogenous, like a blacksmith's tools are, it is not self-replicating.

Thus synthesizers could have the advantages of self-replication, including very low cost, while completely avoiding any gray goo (runaway replicator) possibilities.

Even so, the availability of virtually free home manufacturing has been prophesied by some to imply the breakdown of the economic system. A classic in this vein is the *Venus Equilateral* series by George O. Smith, a radio engineer in the 1940s. Who will work when they can get everything they need from their machine?

The question can be answered by asking its opposite: who needs to work if everyone can get what they need from the machine? Even if people wanted to work, what can they do?

They can invent new stuff to he produced by the synthesizers, for one thing. This does not have to be rocket science, it can just be new styles in clothing. They can maintain the infrastructure that brings the water, gas, and power to the machines. They can do sports, arts, entertainment, politics, law, develop software, discover science, even write books about nanotechnology. Only 30% of our current economy is manufacturing or farming. The rest is information handling and services.

There will always be things (like land, historical objects, and people's attention) that can't be synthesized. People will work and trade to get them. But with synthesizers, the amount they will have to work to have a comfortable life will be much smaller than now.

The major problem with synthesizers is going to be annoyances we are already familiar with today on our computers. That spam offering to sell you X-rated pictures will offer to make sex toys for you in the privacy of your own home. The virus that takes over your computer and scrambles your files could poison your food instead.

Companies will offer free designs for things that work for a month and then fall apart, just as demo versions of software do now. Then they'll want a monthly fee for the working version instead of selling you a real item once and for all. They will try to sell you objects that sound useful at first but for which you have to purchase numerous expensive options for full functionality.

9

NANO-RUNAWAY REPLICATORS

Sometime in the twenty-first century someone will release an aerovore into the Earth's atmosphere. An *aerovore* is a form of *gray goo*, a self-replicating nanomachine that feeds on the natural environment and exists only to convert as much of the universe as possible into copies of itself. The term *aerovore* means a creature that eats air.

Living off thin air is hard work. Life flourishes almost everywhere on Earth, living on rocks and deep sea vents and so forth. No natural life-form (to the author's knowledge) lives on air alone, although the air is full of spores and microorganisms going from one place to another. *Nanotechnology* makes it possible, however, to live on just air and sunlight.

An aerovore would work something like this: It floats in the air like a particle of smoke. Its structure is mostly carbon, which it separates out of the carbon dioxide (CO_2) in the air. It needs energy to do the separation, so it has solar collectors. The solar collectors' are like antennas that stick out the sides, making it look something like a miniature black snowflake. (It would be too small to see with the naked eye.) It would be black to catch the sunlight; some would be converted into power, the rest would warm up the air around it, providing lift. (The aerovore might also use, or consist of, a hydrogen—filled balloon.)

Such a device might be created as a doomsday weapon, for example. The intent would be to have aerovores multiply throughout the atmosphere, causing a nuclear winter effect. Alternatively, an environmentalist group might try to use them to undo the greenhouse effect and stop global warming. We have plenty of evidence that people with either of these motivations can command significant resources and have no compunctions against acting unilaterally.

What will happen when someone releases one? Probably nothing much. An *aerovore* must be a full-fledged *Stage V nanomechanism*, highly optimized. It is trying to live in a niche that natural life can't manage to exploit, after all. It will be able to carry little radiation protection, so it will need to be small and fast to reproduce before it gets zapped by cosmic rays. If it lives very high in the atmosphere, it gets exposed to more radiation. If it lives low, it gets rained out. The aerovore's design is a dense web of engineering trade-offs. It needs lots of power to separate carbon and hydrogen, so it must have plenty of solar collectors. It needs to sort through about three thousand oxygen, nitrogen, and argon molecules for every usually one of (CO_2), so it must have many molecule sorters. It needs to build offspring as fast as possible, so it must be packed with construction machinery. All these needs conflict, and the chances are *aerovores* won't multiply very well if they are not balanced just so.

What's more, building *nanomachines* purely out of just carbon, hydrogen, oxygen, and nitrogen (CHON for short) is a tough constraint. All of the working parts seen in simulation so far use other elements, ranging from sulfur to flourine, to have a better-stocked toolkit of atomic sizes to get the shapes you need. There is a good chance we are going to need metals for various mechanisms. The molecular machines in your cells need a wide variety of metals in trace amounts. An example is the iron in hemoglobin, which gives a flavourable electron distribution to an oxygen-catching pocket in the molecule. The aerovore needs oxygen-catching pockets in its molecular sorting rotors.

Let's assume, however, that the engineering has been gotten right and someone builds an *aerovore* that works. Then what? Do they block out the sun?

The first thing to note is that there is not all that much CO_2 in the air to begin with. Let's assume that the aerovore is a cubic micron and contains ten billion carbon atoms. Building the *aerovore* compacts the carbon from its original form, where it occupies one part in three thousand of the air, to one part in a million. (These are all parts by volume—parts by weight remain unchanged.) It works out to about a thousand *aerovores* per cubic millimeter of air.

Unfortunately, the collectors increase the effective diameter of the *aerovore* to, say, 5 microns. If they convert all the CO_2 the *aerovores* in one cubic meter of air would have an effective surface area of four hundred square meters. Not much sunlight would get through. The Earth, seen from space, would soon look dark, instead of blue-whitc. *Aerovores* killed by radiation would fall to the surface as a fine black snow. Vegetation would begin to die.

Of course, *aerovores* are also a form of vegetation, artificial of course, but dependent on sunlight no less than trees or grass. What's more, they would rapidly eat out their food stock, the CO_2 in the air, at the highest level they could operate optimally (lower levels would be shaded by the higher levels). With low CO_2 levels, aerovores would not be able to replicate fast enough to beat the effects of radiation and would begin to die off. In the long run, they might simply starve themselves out of the air. No one has clone a serious climatological simulation yet, since the parameters of aerovores are not really foreseeable at this point. But it seems possible that they could deplete the CO_2 and largely die off plunging the Earth into an ice age. Or an ice age might he avoided, since black *aerovore* snow might prevent glaciers from reflecting so much sunlight back into space.

NUCLEAR WINTER AND SKYSWEEPERS

Nuclear winters, uninvited ice ages, smudgey skies—all are best avoided if possible. One obvious thing is that aerovores, like any form of vegetation, represent a resource. They, like green plants, have already done the hard work, and we need only harvest the fuel their carbon represents in concentrated form. Rather than digging coal mines in the ground, we need only fly through the air with big filters and catch the stuff. A *nanoengineered* reaper could look like a gauze veil the size of a football field, and float through the air sweeping the sky. (It would be essentially invisible from the ground.) Such a *skysweeper* would burn part of its catch for its own energy and send the rest down to fuel collection depots on the ground. Unlike aerovores, skysweepers could operate at night.

As an aside, aerovore smoke at its densest could fuel an airplane flying through it, but we would not want to let it get that dense—all the bad effects mentioned above would occur.

Skysweepers are a good idea whether *aerovores* are around or not. The sky is frill of all sorts of gunk. You may be surprised to learn that about 10% less sunlight reaches the ground today than in 1950. Some of the haze causing this is from industrial effluents, but more is from biomass burning, for example, wood and dung, that forms a major energy source in the (densely populated) unindustrialized countries. Wherever the gunk comes from, the air could stand to be cleaned up. If we did clean up the atmosphere, we could tolerate a 10% blockage owing to aerovores and still get the same amount of sunlight we do today. Better yet would be to keep it completely clean.

Biovorous Nanoreplicators

The *gray goo* scenario is more often imagined as a form of mechanical bacteria that live on the biosphere. It is commonly imagined by thriller novelists that such *nanoreplicators* could be much more voracious than natural bacteria, reducing the biosphere to gray goo (hence the name) in the timeframe of a major motion picture. On the other hand, other commentators pooh-pooh the idea that such replicators are even possible. Apparently they have not heard of bacteria. But of course the existence of bacteria does not prove the possibility of something significantly more efficient than bacteria to pose a threat.

Birds Fly

We can build flying machines that, operating on basically the same principles, are faster. Yet our airplanes are distinctly inferior to birds in some very important respects. They need a substantial support infrastructure; birds can operate an Airplanes need special highly refined

Fig. 9.1. Flying bird.

fuels and are built of special, highly processed materials; birds get both from the natural environment by their own efforts. Birds repair themselves and reproduce themselves. If you tried to build an airplane that drilled and refined its own fuel, mined, smelted, and shaped its own aluminum, contained its own repair shop, and manufactured new airplanes as well, it is not so clear that it would fly faster than a bird. In fact, it probably could not get off the ground at all.

But bacteria also build themselves out of found materials and use them for fuel. They grow, repair themselves, and replicate. And they do it with remarkably high efficiency.

Nanomachines, on the other hand, will be more like the airplane. They will be made of materials requiring a lot of processing and use specially prepared fuels and feedstocks. They will use higher energy reactions and operate at higher speeds than life's mechanisms. They will be faster than bacteria, just as airplanes are faster than birds. But they have the same drawbacks as well, and for the same reasons.

A novelist can put birds and airplanes together and imagine a creature that lives on worms and flies at the speed of sound. Novelists can also put bacteria and nanomachines together and imagine super-fast nanoreplicators that live on anything and eat the environment out from under us. But in both cases, there just ain't no such animal.

Ostrich Strategy

It is certain that some people reading this prediction will have the immediate reaction that it would be better if *nanotechnology* were never developed at all. Some people are making public cases for this idea already. They would like a law to be passed that would prevent *nanotechnology* from being possible. Their like has been seen before. In 1897, House Bill #246 in the Indiana House of Representatives proclaimed that the value of pi would thenceforth be not its actual irrational value, but one of three different; rational values depending on whether you were trying to find the circumference of 'a circle, its area, or do trigonometry. Luckily for Hoosiers, the bill died in the state senate.

It seems likely that similar things will be attempted for *nano-technology*. Current-day legislators are no more scientifically astute than their nineteenth-century counterparts. For example, Caroline Lucas, the Green Party Member of the European Parliament for southeast England, proclaimed, "The commercial value of *nanotech stems* from the simple fact that the laws of physics do not apply at the molecular level.

Activists and politicians somewhat more in contact with reality will most likely opt for laws that, while implicitly admitting that nanotechnology is possible, will attempt to prevent people from developing or deploying it. That is, they will first stop people who are currently claiming to work on nanotechnology. This will bring a great buzzword shift, as all the people who had been working on chemistry, or surface physics, or coatings, or material science, and began calling it nanotechnology in the past decade will go back to calling it chemistry, or surface physics, or coatings, or materials science. Next, the laws will have to clamp down on all those people and somehow erase all the knowledge they have generated. Similarly, biotechnology, one of the most direct pathways to nanotechnology, will have to be stopped: not just prevented from selling products, but the knowledge curtailed. Finally, you'd have to outlaw high-precision machining and watch making, since the original pathway proposed by Feynman lay in that direction. Indeed, you'd have to stop most technological progress in its tracks, since any improvement in precision control, or chemical synthesis is a step toward nanotechnology.

Well look, they will say. We did it with nuclear technology. Since World War II, nuclear technology has been "*born secret*" in this country. Under the strictest regulation possible there is a death penalty, in some cases, for sharing scientific knowledge, which has been enforced progress of nuclear technology has been effectively quashed. Compared with other technologies in the latter twentieth century, nuclear has been at a standstill. Worldwide treaties prevent proliferation and any development of nuclear technology in heretofore nonnuclear nations is viewed with concern.

And yet proliferation has occurred anyway. It is important to point out that nuclear technology is quite difficult. Whatever path you take has to go through uranium. Uranium is scarce, difficult to work with, easy to detect—and you need tons of it. Even with no restrictions, nuclear technology would be the province of government and large corporations.

Nanotechnology, on the other hand, can be done, if you follow the biotech path, with bacteria that are literally sitting on the end of your nose. You can choose any of hundreds of pathways. The size of an effort to develop it, given current knowledge, is probably about that of the science oriented departments of a typical state university, and getting smaller every day.

What's more, nanotechnology has a lot bigger market than nuclear weapons. Again, only governments can afford to own nukes, but almost everybody has a use for nanotech products. Cancer cure, anyone? Life extension? Intelligence enhancers, fat reducers, AIDS prevention?

Another major difference is that nuclear technology was throttled without strangling a lot of existing, economically important, fields. Things nuclear are a world apart from the operations of ordinary matter. But life itself—medicine, understanding ourselves and our environment—is one of the major pathways to nanotechnology. Mechanical engineering is another catalytic chemistry, materials science, microchip fabrication: virtually all the technical background of standard products today forms a thicket through which a host of pathways could be picked.

If general laws are passed to impede *nanotech development*, the military in the United States and almost every other country will be exempt and develop it anyway. So will organized crime, various international drug cartels, and probably some environmental groups. Luddite movements will attempt to demonize nanotechnology as well as other technologies, but people will start smelling a rat when senators continue to hold office past age 120.

Short of plunging the whole world into a Dark Ages, you simply can't hide the laws of physics. The basic physics necessary for nanotechnology has been in the textbooks for decades. The ideas are out there, the pathways are many. The more that legitimate, open, advanced technology is restricted, the more people will look for other options.

An old saw says that if the government really wanted to help literacy and reduce addiction in the inner cities, it would form a Department of Drugs and declare a War on Education. It is a fact that in random tests of airport security procedures after the tightening up that followed September 11, 2001, roughly half of the test weapons got through undetected. Grand government prohibitions are primarily a security blanket and provide only a false sense of safety.

Laws that do work, more or less, such as those against murder, work because almost everybody agrees with them, only a very few people try to break them, and those people are chased and mostly caught after they do break the law. Note the "after" part.

The more people want what is prohibited, the bigger the black market will be. Drugs are impossible to stop, but are seriously sought by a relative few. A better model for nanotech benefits might be Prohibition, when a major segment of the population was in the market. The laws were essentially laughed at. And yet some people believe that technological progress, and all its enormous benefits to virtually everybody, can be stopped by similar restrictions. These people are simply in denial.

With *nanotechnology*, the stakes are much higher. With a self-replicating threat, the enforcement web need miss only one development effort in the entire world to fail. And when that happens, the aerovores will be there, but we will have prevented any development of skysweepers. It is too great a risk to take. Indeed, it is not really a risk but a certainty: if open, aboveboard nanotechnology development is stopped or even seriously delayed, calamity is assured.

Paint, Soap and Washing Machines

If, on the other hand, *nanotechnology* is developed evenly so that the mainstream, in general has the same level of capability as rogue states and rogue groups, there's no great problem. Nanotech threats that would be catastrophic to a world equipped only with conventional bulk technology will instead be simply spills to be mopped up, fires to be put out, and generally preventable by ordinary hygiene and maintenance.

Imagine that it is 1900. Some people (like the Wright brothers) believe heavier-that-air flight is possible, and others do not. Some people have become alarmed by the idea, and they spin out a scenario in which a group of bandits build a plane, at a level of, say, a WWII bomber such as a B-29. In 1900 such a group could have leveled Washington without resistance and dictated terms to the country.

An *alarmist* at that date might well have drawn such a picture and urged that heavier-than-air research be banned. People actually building kites and wood-and-silk gliders would have scoffed at the proposed capabilities of a B-29, and the scoffing would be well with the common sense of the day. That alarmism and scoffing can clearly be seen in public discussions of nanotechnology today. Neither side has a handle on the truth.

By 1950 a rogue group with a single B-29 would have been blown out of the sky by jet fighters. Yet equal airplanes plied the air every day carrying passengers and freight. No one worried about some criminal group buying an airliner and using it as a bomber, because air forces were in place and the opportunity for major mischief was gone. Criminal groups had much more profitable things to spend their money on.

The notion that someone could have built a B-29 in 1900 and wielded it against an unsuspecting world is ludicrous. So are the scenarios you hear about hill-fledged *Stage V nanosystems* loosed on today's world. Technology imagined out of temporal context sounds a lot scarier than it actually is. Let's take another example, this time not of evildoing, but of an accident.

Used tires are a major problem. They are hard to get rid of, they don't recycle easily, and they accumulate in giant piles that sometimes catch fire. (The fires are quite dangerous and difficult to put out.) Suppose some bright person decides to convert this problem into a resource. He invents a tire rot that lives on tire rubber and, using some of it for energy, converts the rest into oil. Then you take a big oil storage tank and start throwing tires into it, and first thing you know, you have gotten rid of the tires and have a full tank.

You know what happens next. Some of the rot escapes and finds its way onto a tire on a vehicle. The tire starts to rot away, leaving a trail of oil all full of *rot nanobots* on the road. By the time the first tire is gone, every road in the state has little oil tracks and anyone who drives there picks it up. In a week it is all over the country; and cars and trucks are worthless.

That's what happens if we assume nanotechnology completely out of its own time, or a situation where mainstream nanotechnology has been suppressed. There are several much more realistic scenarios, in a world where nanotech development is embraced by the mainstream:

1. The *inventor*, trained in engineering schools where safe design was taught, and using widely available tools, finch it easier, as well as the only legal way, to design a system with built—in limitations and safeguards.
2. The *new rot* is quickly detected by the ubiquitous environmental sensor net. It is analyzed, and poison molecules, carefully designed to affect only this one replicator, are released through the network. People in affected areas are warned, they download the recipe for the poison to their home synthesizers, and spray it on their tires.
3. *Rubber rot* is out there already, so tires are built with immune systems. At a decent level of nanotechnology; it is stupid and wasteful to build a tire as a balloon of bulk material anyway. Instead, they are non wearing, self-repairing machines that literally grab the road at a microscopic level—a lot smoother riding and safer than balloon tires.
4. Everybody uses *hovercraft* or flying cars by now, so nobody cares. The last of the old rubber tires are eaten away, and the spots are sprayed with standard oil-slick-removal nanobots.

The more advanced the general nanotech level of society in general is, the less a problem the release—which would be a disaster today—would amount to.

Wood Rots

If you take a piece of untreated wood and set it on the ground in a natural environment, it will be spongy and crumbly within a year. If you cut your finger and don't apply antiseptic, you will soon find that microscopic machines—natural ones, bacteria—are waiting everywhere to munch on you, as well. New natural munchers evolve all the time. Simple poisons we have developed over the last century, antibiotics, go out of date as the organisms evolve around them.

The nanotech equivalent of paint for wood and soap for our bodies is clearly something we will be better off developing no matter what. By the time new mechanical munchers emerge, we will be perfectly ready for them, if in the meantime we have not been foolishly ignorant, or panicked with our heads in the sand.

10

Environmental Nanoscience

It is important to remember that both security and nanotechnology are complex subjects. The threats to societal security come in many guises, from something as simple (if frightening) as a bomb attack on an office building to more subtle means such as disseminating computer viruses or electronic disinformation or tampering with drug packaging and distribution.

Security concerns go well beyond those associated with terrorism, ranging from new diseases like AIDS, SARS, and Ebola to the tremendous challenges associated with climate change, energy supplies, and global warming. *Nanotechnology* can be of great value in addressing some of these concerns, but others extend into realms beyond the scope of physical technologies.

To take one simple example: If there were no motivation for acts of terrorism, there would be no acts of terror. These motivations can come from economic inequity and uncertainty, from religious belief or incitement, from ethnic or religious hatreds and traditions. They may even come from a combination of factors featuring civic pride—every Super Bowl win or NBA championship might well start looting, rioting, and burning in the winner's city.

All technology (including nanotechnology) can change the economic, education, environmental, and employment patterns in society. This change can cause concerns about everything from globalization to job security to environmental dangers. These concerns extend well beyond the scope of our discussions, but are a major aspect of terrorism and of homeland defense.

Fabrication

Nearly all manufacturing processes have substantial environmental impact. At the height of the industrial revolution in England, streams were polluted with animal wastes and chemical by-products, and the skies were often thick with killer smogs of particulate carbon smoke and sulfur dioxide. In Silicon Valley, several of the by products of high tech *Fabrications* have created large environmental problems. Almost no American city has drinking supplies that are clean enough to use without treatment, and brown field and Superfund sites (industrial areas that were left polluted by a previous owner) are becoming more and more abundant. Anyone who has lived near a steel mill, oil well, tannery, foundry, or paper plant can vouch for the unpleasant environmental aspects of manufacturing.

There is a great deal of concern that nano scale manufacturing could augment these problems. The reality is probably just the opposite. *Fabrication of nanostructures* is in many ways less hazardous than manufacture of more traditional industrial products because the control must necessarily be much better. In an *automotive assembly line*, for example, welding is often done in the open and little can be done to contain waste gases, so these are simply vented. Raw materials for macro scale manufacturing often include nasty agents like chlorine or the hydrogen cyanide (HCN) gas we talked about as a chemical weapon. In the manufacture of nanostructures, however, not only are the amounts of material used typically smaller and the components more benign, but the precision and chemical sensitivity of the manufacturing processes require that; fabrication be done in closely controlled systems under laboratory conditions. This makes waste easier to capture and, once captured easier to treat.

Still, *nanoscale manufacturing* involves dealing with materials that could have harmful effects very different from those used in traditional manufacturing. Consider biological nanostructures, for example. The design of sensors, improved drugs, and a variety of nanotechnology-powered devices involves using DNA, bacteria, and viruses. Some people worry that biologically active agents might be released into the environment where they could interact with other biostructures (including plants, animals, or people), perhaps leading to substantial new health hazards. This concern is similar to the worry about genetically modified organisms in food, or (in an earlier time) concerns about medical waste and hybrid farm crops including corn and wheat.

With both biological structures and their non biological counterparts, the fundamental issue in limiting environmental damage is controlling the effluent streams from manufacturing or health facilities. In a fabrication plant for *biological nanostructures*, this task should be fairly straightforward, since most operations would probably be carried out by robots in controlled environment chambers (analogous to a semiconductor plant's clean room). Air coming into and out of the chamber can be filtered, purged, and irradiated so that no biologically viable products are released into the environment. The few workers who do enter can wear nano-enhanced defense suits, and outside surfaces of chambers and equipment could have biocidal coatings.

Of course it is true that these filtration systems might break down, but because the volumes involved would be relatively small and the manufacturing processes very highly controlled, the rate of emission of viable biological material into the environment could be made essentially negligible. Even if such materials were to escape, nearly all of them would break down rapidly—both simple DNA strands and proteins denature quickly when exposed to the outside environment. Even if the scale of manufacture or the extent to which people are directly involved is greater than expected there are established procedures for running these types of plants safely. After all, the problem of containing and treating biological entities used in manufacturing is solved every day in the world's pharmaceutical factories.

Another concern raised by activist organizations like the ETC Group is that even non biological nanostructures (which one might expect to be benign) can in fact be highly toxic. There are a few reasons for this. One is the extreme reactivity of nano scale materials (the same property that makes them so good as catalysts), and the other is that these ultra-fine particles are the right size to be absorbed by cells and membranes in much the same way that asbestos is. While sheets of asbestos or carbon are not particularly toxic, tiny airborne particles of either could be trapped in the respiratory tract and become extremely carcinogenic. In the manufacturing context, this problem can easily be addressed using the same techniques we've just discussed. However, more work is required to deter mine how great these risks really are, especially for manufacturing such products as skin lotion, medicine, or other material to be used on people. Absorption into the body is a potential issue only for certain classes of nano structures. Integrated nano devices, where a nanostructure is physically integrated into a larger device such as a chip, smart materials, and most other developments from nanotechnology, will be immune.

In sum, the fabrication of nanostructures has some unique challenges but nanoscience also provides solutions to most of them. Additional regulatory oversight for nanomanufacturers is necessary, but on balance nanofabrication should produce less waste and few new environmental hazards if it is well managed. Such management is well worth the effort, since the next few decades have the potential to be the best postindustrial period yet for the Earth itself and for the people on it.

Remediation of Ongoing Environmental Issues

Environmental awareness has been one of the most striking, impressive, and important political changes in the past half century. Although environmental activism has roots going back to Thoreau and beyond, the naturalists, writers, sportsmen, artists, political figures, and social leaders who brought about the modern environmental awareness revolution have contributed tremendously to society's evaluation of itself and perhaps to its very survival. Their contributions create the possibility for living in a friendlier, safer, more beautiful, and more rewarding world.

While concern about the environment is abundantly justified, nanoscience challenges the common activist argument that industrial and economic growth are necessarily incompatible with sound environmental management. Many ongoing environmental issues have been caused by development and industrial progress without appropriate environmental considerations, but nanotechnology could offer cleaner ways to make things, as well as substantial help in remediating existing environmental hazards.

The three main areas of environmental concern are all earth, and water. Both air and water can be filtered to remove toxins and pollutants. Because many common pollutants are composed of particles that are themselves nano sized, nanoscale filters are ideal for purification. Such ultrafiltration methods are already common in industrial areas, where they have proven to be relatively cheap and very effective, but new nanotechnological developments should permit these filters to remove many new types of waste products and to be longer lived, many times more efficient, and capable of handling much larger volume. It is difficult to imagine cleaning a large polluted lake or a city air shed by filtration—how do you get that much air or water through a filter? However, filtration or treatment closer to the source, as with catalytic converters on cars and sewage treatment facilities, can result in the reduction or elimination of pollutants before they ever get into the environment. If these devices are adopted (and incentives or regulation may be required for this to happen), large-scale post pollution remediation will not be as necessary in the future as it has been in the past. Once the rate of new pollution diminishes, the Earth has shown that it has a wonderful ability to heal itself. Many of the processes it uses involve natural nanotechnology.

In some cases, though, the natural rate at which the earth heals itself is not sufficient. The Valdez oil spill and a chemical weapons attack are obvious examples. In these cases a more aggressive approach to renditions may be necessary. Although there is no universal solution, nanotechnology is providing approaches for dealing with some of the more common and aggressive pollutants. For example, particular ions such as chromate or arsenate can end up in water supplies and are extremely carcinogenic. Ions cannot be easily filtered, but a simple chemical exchange can replace these toxic ions with more benign alternatives such as ferrous iron. While this has been known for some time, the efficiency of this replacement process is limited by the amount of surface area of the reactants. As the particles of treatment chemicals are reduced to the nano scale, they essentially become all surface. This means that the process of replacing toxic ions with benign ones can be greatly sped up, made much more efficient, and made much cheaper.

The capability of nanostructures to sense certain molecules could also be useful for dealing with a number of important environmental problems, especially those that result from military campaigns. For example, one of the nastiest residues of military conflict is unexploded ordnance, including bombs, grenades, and landmines that can remain hidden on the battlefield and cause danger, fear, injury, and death long after the war is over. Sometimes this ordnance can be destroyed using bulldozers and earthmovers, but rocky or mountainous terrain can make that virtually impossible. Even human mine clearing is not always 100% effective (though it is nearly 100% risky), and in places like the Middle East there are large tracts of land that are only approved for animal pasturage since no amount of clearing can guarantee that all the mines are gone. This isn't a problem limited to the developing world. There are still unexploded shells from World War II on Salisbury Plain in England and in Danish coastal waters.

Until recently, soldiers looked for mines using metal detectors. Military technology kept pace, however, and mines now exist that are mostly plastic and thus virtually impossible to find with a metal detector. Chemical sensing provides an alternative. Since nearly all unexploded ordnance contains nitrate propellants and since nitrates are relatively easily sensed (the white fabrics used in airports to swab luggage do just this), nanoscale sensors may be able to find this unexploded ordnance even if it is buried deep in the ground. Removing the scourge of mines and bombs from the fields in which they are strewn would be of tremendous value in returning the earth to its productive and rich self. Finding hidden weapons is also crucial for dismantling the military apparatus of a foe once a battle is over, and it would be a boon to weapons inspectors trying to prevent a war from happening in the first place.

Public health depends upon the clean environment, but certain structures, notably microorganisms, can affect health even in very small amounts. Nanoscience, through the development of integrated sensors, offers the possibility of total monitoring for public health applications. Such total monitoring is not the invasion of personal privacy as implied by its name. Rather, it means that air, water, and soil could be continuously monitored for the appearance of any known toxic agents. Such monitoring could allow instantaneous response to chemical leaks, filter failures, industrial accidents, and, of course, bio- or eco-terrorism.

A combination of these techniques—sensing, nanocatalysis, ion replacement nano filtration, pervasive computing, and others yet undeveloped—can mightily increase the power of the tools we have to fight pollution and clean up our environment. Some of them are in use already. We hope more will be deployed in the next few years.

Transportation

Transportation is one of the major polluters of the air. Estimates for the total contribution of transportation to the air pollution burden in the developing countries run up to 50%. The results of this air pollution are obvious in big cities. Replacing diesel or internal combustion engines with portable electric power provided by fuel cells or batteries would cut the emissions of carbon monoxide, particulates nitric oxides, and sulfur oxides to essentially zero for all cars, trucks, and trains powered by these alternative means.

The idea of using electric powered cars goes back to the beginning of the century (some readers may remember Grandma Duck's Detroit Electric, built in Michigan in 1908), but as of 2004 there are still very few electric powered vehicles on the roads. There are a few reasons for this. Early electric vehicles could not match the acceleration or performance of their fossil-fueled cousins, regulatory pressure such as California's zero emission plan and federal clean air guidelines have been scrapped under pressure from corporate interests, and not enough research has been done to make fuel cells and batteries efficient enough to compete on their own.

Fortunately, the situation is beginning to change. Fuel-conscious buyers ranging from religious groups to corporate fleets are pressuring manufacturers for ecologically responsible vehicles. Also, slowing sales of conventional vehicles are forcing automakers to look for the next big thing. Investment in fuel cell research in particular has soared in 2003. Hybrid vehicles, which use a combination of fossil fuel (for rapid acceleration) and electric power, are gaining some popularity already, and they are a very attractive medium-term solution. Nanotechnology is promising to improve both batteries and fuel cells to the point where they may become the sole power source for a car or truck without huge performance sacrifices.

Improving batteries and fuel cells can be done in a number of ways. The key is to improve the power source by making the chemical reactions at the electrodes and the motion of ions within the electrolytes faster and more efficient. This translates both into higher energy density (energy to weight ratio) for longer running time between recharges or refueling and into higher power for better acceleration. Interestingly, some of the most exciting fuel cell research is happening in Texas despite that state's considerable stake in the fossil fuel economy.

Efficient fuel cells for other purposes already exist. NEC has announced a laptop computer based on a methanol fuel cell that it claims can run for five hours, and a 40-hour version should ship by 2005. This means that these fuel cells will have ten times the performance of the best lithium ion batteries—and they're just getting started. The fuel cell industry is still in its infancy, and when it matures to the point that batteries have, the results will be staggering.

Fuel cells work by extracting energy from a chemical reaction rather than by burning fuel like an internal-combustion engine or storing energy like a battery. This means that spent reactants can be captured within the cell rather than vented off (as in the exhaust system of a car) so they can be disposed of cleanly. While different designs are necessary to use different kinds of fuel, successful fuel cells have already been run on a variety of fuels including methanol, pure hydrogen, and methane. Methanol is a common additive in gasohol and other fuels since it is relatively easy to make from natural gas, compost, or other waste products. This makes it a very attractive candidate in the short term, though other fuels have better long-term potential. One of the other options, natural gas or methane, is a common waste gas from landfills. (The fires you see at night on most landfills are caused by vented methane being burned off.) The use of methane means that fuel cells could be installed at landfill sites, where they would be powered by products normally considered to be waste.

While batteries don't generate energy the same way fuel cells do, they allow us to use electricity generated at industrial power stations, which are much more efficient in converting fossil fuels to energy than current cars are. Batteries also eliminate the need for a gasoline distribution system. Using either of these alternative energy sources would permit great reduction in air pollution load directly from transportation.

ENERGY

The supply of energy is probably the most important physical problem that civilization now faces. As Richard Smalley has pointed out, there is no driving economic or environmental problem on Earth that will not be alleviated or at least reduced if a source of readily available, clean, and inexpensive energy could be developed. Energy powers industry, heats homes, and moves goods. It can also be used to separate salt from seawater, to grow crops, and to eliminate drought and hunger.

Clean energy would have an overwhelmingly positive effect on the environment. It could reduce, eliminate, or potentially reverse global warming and climate change. These are the two great environmental time bombs. If they are not stopped, they will cause great harm to our society, resulting in everything from flooded coastlines and melted icecaps to changed rainfall pat terns that produce deserts where farms once existed. The amount of carbon dioxide (one of the important greenhouse gases that causes global warming) in the atom sphere has gone up more in the last hundred years than in the previous half million years. The reason for this is simply more human life on the planet leading to more industrial production, more breathing, more burning, and thus more carbon dioxide.

The solution to global warming, as was recognized in the Rio and Kyoto Agreements, lies principally in the reduction of carbon dioxide emission into the atmosphere. Almost all carbon dioxide is produced by burning carbon-containing fuels. Animals breathe out carbon dioxide because they burn sugars that contain carbon (though respiration

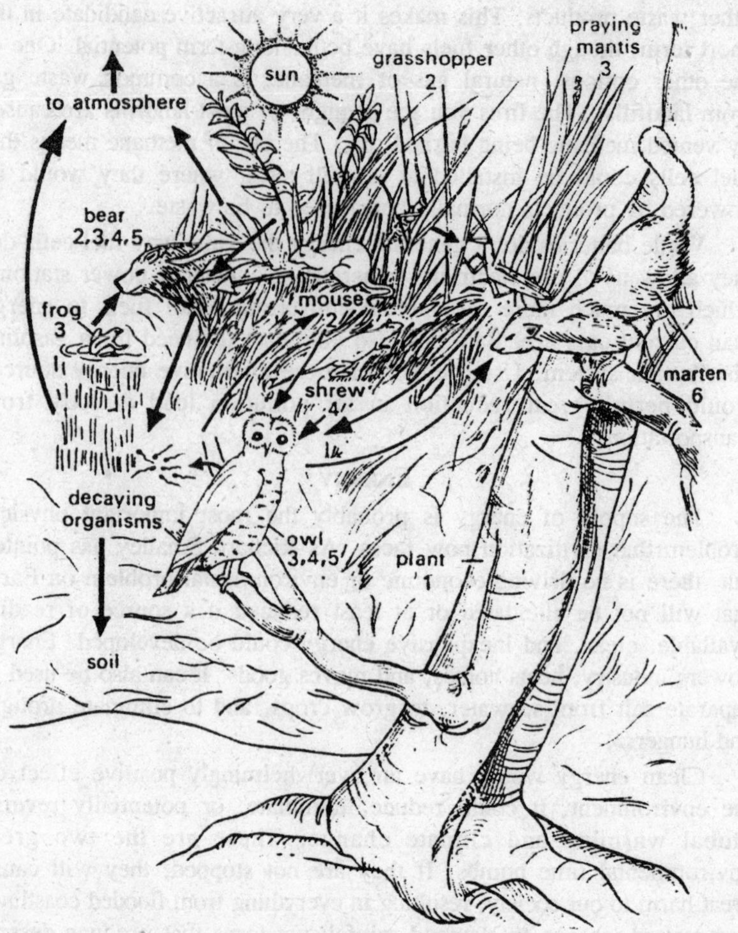

Fig. 10.1. Flow of energy.

constitutes a tiny fraction of carbon dioxide emission worldwide). Fires send out carbon dioxide because we burn wood, coal, oil, peat, or paper, all of which contain carbon. Greenhouse gases like carbon dioxide are a threat because when they are released into the atmosphere they block the escape of the heat caused by sunlight shining on Earth, effectively making the earth into a greenhouse or reflector oven. This is the major source of global warming, and one that we will have to control if society as we know it is to survive.

Dealing with the *carbon dioxide* load and global warming will require the development of energy sources that do not involve combustion

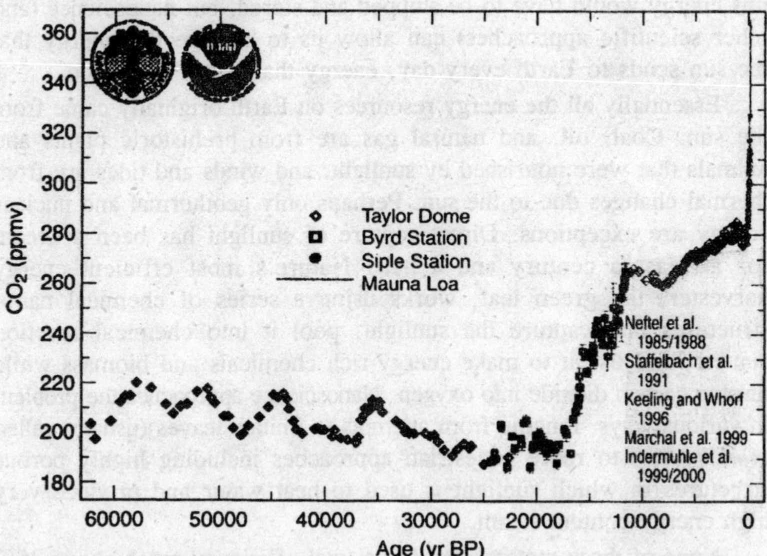

Fig. 10.2. The image shows carbon dioxide (CO_2) concentraion in the atmosphere over the last 62,000 years.

of hydrocarbons. Petroleum should be used mostly for its capability to make materials such as plastics or pigments, rather than for burning. This would also have substantial strategic advantages since it would make the United States independent of foreign oil. Domestic petroleum supplies are more than ample for non fuel purposes.

Carbon-free sources of energy include wind, geothermal, tidal, nuclear fission, nuclear fusion, and solar energy. While conservation is not strictly a source of energy, more efficient use of existing energy resources and reduction of unnecessary use also help relieve the strain. All of these avenues are important, and all of them are being pursued. Nanoscience and nanotechnology can help substantially in all of them, but perhaps most obviously in conservation, in the development of solar energy, and in improving the efficiency of existing devices.

Solar Energy

Let's start by looking at *solar energy*. If we could design nanostructures capable of capturing sunlight with an efficiency of 50 percent, then the entire amount of energy required to run every home, business, car, truck, and toy in the United States could be captured by erecting solar energy cells on only a 40 mile by 40 mile area of land in the Nevada, Arizona, or New Mexico desert (based on projections from the National Renew able Energy Laboratory [NREL]. Of course this energy would have to be shipped and stored, but nanoscience (and other scientific approaches) can allow us to capture the energy that the sun sends to Earth every day, energy that is not now used.

Essentially all the energy resources on Earth originally came from the sun. Coal, oil, and natural gas are from prehistoric plants and animals that were nourished by sunlight, and winds and tides are from thermal changes due to the sun. Perhaps only geothermal and nuclear energy are exceptions. Direct capture of sunlight has been a dream for at least a century and a half. Nature's most efficient energy harvester, the green leaf, works using a series of chemical nanostructures that capture the sunlight, pool it into chemical reaction centers, and use it to make energy-rich chemicals and biomass while turning carbon dioxide into oxygen. Nanoscience approaches the problem in various ways, ranging from attempts to imitate leaves (usually called *biomimetics*) to more pedestrian approaches including highly porous structures in which sunlight is used to heat water and produce very high energy content steam.

None of these methods yet has a total efficiency much above 25% (though this is up from 3% just a few years ago), but through continued engineering and development we can anticipate that large-scale solar capture will become real and efficient. Startups like Nanosolar are already making better nano enhanced solar collectors. Not only are these new cells more efficient, but Nano solar claims that they are flexible, so they can be wrapped over any surface. Moreover, they are only 300 nanometers thick—around 1000 times thinner than the average silicon-based solar cell in use today.

Nanotechnology for energy conservation is also an area of active research. One promising approach works very differently than most conservation methods because it doesn't target usage in businesses or at home. Instead, it targets the energy distribution network—the high voltage wires that transmit energy from generation plants to where it is actually used. These wires waste a vast amount of energy—estimates range around 30%—simply because copper wire cannot conduct electricity perfectly. Some 0 the energy is always dissipated as waste heat and the more distance the electricity travels, the more is wasted. Nanotubes offer an intriguing alternative since they can conduct electricity without resistance when properly configured. This means that a nanotube power cable would dissipate little or no power and immediately result in a remarkable reduction in energy needs without altering productive usage. Unfortunately, it will not be practical to make enough nanotubes for such a cable until fabrication technology for nanotubes gets much better.

With improvements like efficient solar power and nearly resistance-free transmission, the world can deal with climate change and global warming, and do so in an atmosphere of energy security. Adequate supplies of energy are needed if society is to permit its members to live lives of comfort and capability, without deleterious environmental impacts. The fact that energy production in the developing world is not yet sufficient to serve its nations' needs is a double-edged sword. On the one hand, it is certainly desirable that these countries should have the energy they need to have good living and working standards. On the other hand, if every country on earth generated energy in the environmentally harmful way that the developed world, now does, climate change effects would be dreadful. Energy remains a global challenge and one of the great problems that we must and will solve.

Solutions like efficient solar power and clean nuclear fusion power may still be far away, but nanotechnology already impacts some aspects of the energy supply situation in an important way. For example, gasoline is made by "*cracking*" natural petroleum. Cracking is a chemical process in which large molecules in petroleum are broken down into smaller molecules, whose octane rating is high enough to be useful as a motor fuel. One of the major advances in the cracking process occurred when scientists at Mobil developed a series of catalysts for making higher-octane gasoline from the same petroleum feed-stock. The trick involved using minerals called *zeolites*, materials made familiar by their use in water softeners. Zeolites are mixed oxides of aluminum and silicon. They have nanoscale pores and chambers throughout their structures that make them look and act like molecular sieves. Cracking using particular zeolites, with appropriate pore sizes and structures, produces much higher octane gasolines. This improvement in making gasoline from petroleum is entirely a matter of nanotechnology because the engineering of the pore sizes and pore crossings to make the best cracking catalyst involves an understanding of the structure at the nanoscale. The contribution of these catalysts to the U.S. economy is billions of dollars per year already. Nanoscience's applications in the energy field have begun.

HYDROGEN ECONOMY

Hydrogen gas is in many ways an ideal fuel. It has very high energy and very high power per pound. It is also a clean fuel, since it burns cleanly with oxygen and produces only pure water as a waste product. Hydrogen can be produced directly from water using electricity or possibly using sunlight directly. Because of these striking advantages, there has been extensive discussion of using hydrogen as the entire basis of energy in our economy—to create a hydrogen economy rather than a coal or oil economy. This hydrogen-based energy model has a great deal to recommend it, including political and security advantages (there is more water in the United States than oil in the Middle East) in addition to the obvious environmental benefits.

In order for the hydrogen economy to be practical we must be able to use, make, distribute, and store hydrogen efficiently. Some of these challenges are easier than others. Using hydrogen is the easy part. We have already looked at fuel cells that can burn hydrogen, and it can also be burned in more conventional turbines. Distributing it is also technically straight for ward since it can be sent through pipelines as a gas or placed in high-pressure vessels (tank cars or cylinders) and moved by truck, train, or ship. Unfortunately, refitting existing systems may be quite expensive.

Producing hydrogen is a bit more complicated. It is best made from water since two of the three atoms in a water molecule are hydrogen. While it can be created from water by electricity in a process called electrolysis, the electricity required for this method has to come from somewhere, and therefore this process largely defeats the purpose unless the electricity comes from another clean source like wind, geothermal energy, or, in the more distant future, nuclear fusion. Other methods for extracting hydrogen are being examined, and intense research is now being devoted to the hard problem of using nano scale methods for making hydrogen from water by direct photo conversion of sunlight.

The hardest nut to crack may well be hydrogen storage. Hydrogen is a gas at ordinary pressure and temperature, and it is very flammable and volatile (the Hindenburg airship disaster arose from the combustion of the hydrogen used to keep the airship afloat). While hydrogen can be stored at high pressure in tanks, the tanks are heavy and it requires a good deal of energy to compress the gas. In most proposed hydrogen storage methods, the storage medium actually weighs a great deal more than the material being stored, making storage of hydrogen quite inefficient on a weight basis. Nanoscience, including biological nanoscience, offers a few possibilities, including storage in carbon powder, nano tubes, or some promising biomaterials. Solving this problem would be a major economic advance.

Domestic Economic Implications

Information *tech*, *biotech*, and *nanotech* all function as technological enablers. They encourage new investment, they produce new goods and services, and they provide new jobs. Because of the widespread applicability of nano processes and nanostructures, nanotechnology could engender an economic boom similar to the one provided by information technology in the 1990s. Even after the first boom it will remain a massive force in the economy.

Drugs can be injected, ingested, or even breathed into the body, but it is difficult to make sure that the active ingredient in the drug gets to the part of the body where it is needed (a tumour or inflamed tissue for example), since most drugs aren't very soluble in water or blood. Most of the medicine actually passes right through the body without being used. The amount that gets to where it is needed compared to the amount consumed represents the bio-availability of a drug. With drugs as expensive as AIDS treatment cocktails (which can run over $ 1.0,000 per year), even a small amount of waste gives a major increase in cost.

While many approaches to the *bio-availability* problem seem reasonable, one of the simplest involves using the nano scale. Because the rate at which a drug dissolves depends on the surface area of the

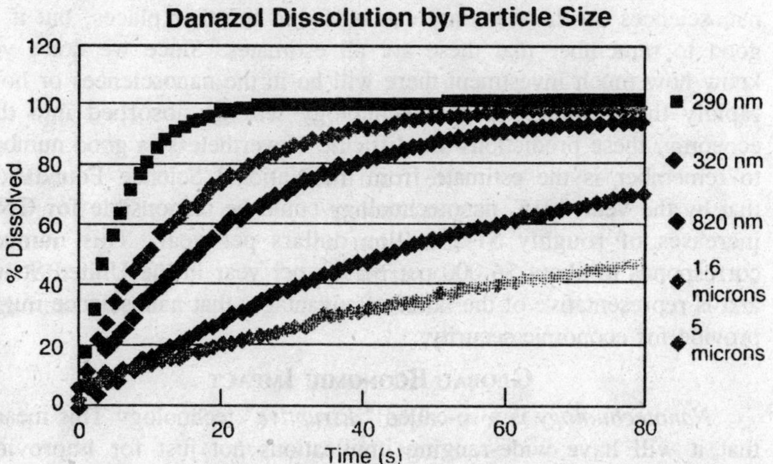

Fig. 10.3. Increasing bioavilability of medicine upon grinding it to nanoscale size.

powder or tablet, using grinding techniques to reduce the powder size to the nano scale will substantially increase solubility and, therefore, bio-availability.

The effectiveness of the drug increased by several times, simply by grinding the particles down to the nano scale. How much impact this new technique will have in solving the bio-availability problem is still unclear, but it is clear that the control offered by nano scale behaviour, even in something so simple as solubility can have major implications in medicine just as it does in manufacturing, transportation, and other important sectors of the economy.

Awareness of the economic possibilities offered by nanoscience is widespread, and is one reason why *Spiderman* and the *X-files* kept *nano* in their scripts. But less televised developments also show how big it will be to get small. Phillip Bond, Undersecretary of Commerce, is bullish on nanoscience, saying, "It is phenomenal what the future could be in this space. It's almost literally beyond most people's imagination. It's going to change everything." The number of nanotech start-ups doubled to more than 1200 in the past year, and is expected to double again. By some estimates more than 5% of all the venture capital investments made in the United States in early 2003 went into nanotech or related fields. Mike Roco, Director of the National Nanotechnology Initiative, has stated that in early 2003 more than 1700 companies in 34 countries were pursuing the commercial possibilities of nanotechnology.

More extended discussions of the possible implications of nanosciences on the economy can be found in other places, but it is good to remember that these are all estimates. Since we don't yet know how much investment there will be in the nanosciences or how rapidly the products of nanotechnology will be absorbed into the economy, these predictions are difficult. Nevertheless, a good number to remember is the estimate from the National Science Foundation that by the year 2015, nanotechnology could be responsible for GNP increases of roughly $1-2 trillion dollars per year. This number corresponds to about $6,000 per person per year in the United States and is representative of the kinds of advantages that nanoscience might provide for economic security.

Global Economic Impact

Nanotechnology is a so-called "*disruptive*" technology This means that it will have wide-ranging implications not just for improving current technologies, but also for creating new kinds of devices and capabilities with ramifications for the very way we live our lives. Previous examples of disruptive technologies include biotechnology, information technology, flight, and electricity. Nanotechnology may be as transforming as any of these.

There is another ramification of many disruptive technologies. Since they are not necessarily built on the foundations of older technologies, they provide opportunities for anyone to get ahead. The story behind this is an interesting illustration of why disruptive technology is so important. Research in bioelectronic computing really began in earnest during the 1980s. Leaders in Russia (still the Soviet Union at that time) feared the United States' dominance in the then emerging field of information technology; and they worried about how to compete head to head in developing the new technology. Instead of putting all of its resources into improving conventional computing, Russia therefore invested heavily in research to develop totally new computing techniques not based on silicon. If it could be the first to develop these, and if their potential outstripped that of silicon electronics (or their manufacture turned out to be easier), the Soviet Union could leapfrog the United States in the information race. The plan didn't work out quickly enough for the Soviet Union—bioelectronics and other architectures proved evasive and are only now becoming reality—but the initiative was strategically sound. Disruptive technology gives little or no advantage to the entrenched leader of an earlier technological wave.

Another example of the power of disruptive technologies is India's leadership in the software sector. India did not start with any natural advantages in developing software when it starting trying to create a software industry in the 1980s and 1990s. It did not have a strong semiconductor business, computers were difficult to import, its telecommunications infra-structure was primitive, and almost all of the basic technology was being created overseas. However, by combining extensive private and public investing with crucial government reforms and strong entrepreneurial leadership, India has outstripped the industrialized economies of Europe and now produces around a third of the world's software. Software has become a key sector of the entire Indian economy and, from an international perspective, probably the most important sector of all.

Nanotechnology could work the same way. China, Japan, Australia, Canada, and most European nations have already set up national nanotechnology initiatives of their own. The governments of Russia,

Israel, and other powerful nations have also established national nanotechnology centers. While the budgets vary widely, some of these efforts (notably China's and Japan's) are comparable in size to America's, and some of the best work in nanotechnology is coming from overseas. Since all of these nations have excellent scientists generating world-class work, the question of which economies (and militaries) will be most affected by nanotechnology is still very much open to contention. The answer to this could have a great effect on the balance of power.

Economics of Security

Economic health and national stability go hand in hand. The poorest nations in the world are also the most troubled—North Korea, most countries of Africa, and most of the nations of the Middle East are prime examples. Economic downturn has led to coups, riots, and even the collapse of the Soviet Union. By its disruptive nature, nanotechnology provides a new set of opportunities for those who embrace it. The potential for it to be an industry generating $6,000 in GNP for every citizen of the United States within 15 years is particularly striking when you consider that this increase alone is much more than the total yearly income of most of the world's people. Many of these will now have the opportunity to use nanotechnology to follow India's software example.

Another important security aspect of nanotechnology lies in its global nature. Thomas Friedman, the New York Times's foreign affairs columnist, has proposed that the tighter economic ties between countries involved in global trade and commerce have resulted in a "golden straitjacket." This straitjacket is a means of gaining great wealth through trade, but it carries with it several constraints including the requirements that trade must remain open and international relations must remain warm. War between two nations with free market economies and truly democratic governments is so rare as to be almost inconceivable. If it continues along its current development path, nanotechnology will continue to grow internationally and become one of the great engines of international commerce. If Friedman is correct, that alone would be a major contribution to international security.

11

Politics of Technology

The Foresight Institute, with which the public is most familiar, is oriented toward educational and informational activities. Its broad task is to make people aware of the concept and possibility of nanotechnology, and to promote discussion on the topic. The Institute for Molecular Manufacturing is the *promoter* and *sponsor* of technical research in molecular nanotechnology. Its function is to increase the likelihood that technical work is performed in this country, and that such research is directed into useful approaches. The Center for Constitutional Issues in Technology (CCIT) is the public-policy-oriented organization of the Foresight Family, and its purpose is specifically to promote discussion on public-policy issues, and (*analogously* to the aim of IMM in the technical-research sphere) to attempt to influence thinking in the public-policy sphere. CCIT wishes to ensure that discussions on the public-policy issues engendered by nanotechnology occur, and that they go in directions that we think are productive.

CCIT was founded in 1991. It has spent the intervening time in organizational tasks, in defining its areas of research, and in preparing potential research programs. Currently CCIT has prepared an initiative regarding a near-term issue—standards in emerging nanotechnology. This program would examine some of the questions regarding product standards in emerging nanotechnology and molecular manufacturing, comparing governmental versus industry roles in emerging standards.

CCIT would also like to examine some long-term issues regarding equitable access to resources, an issue that, as you can tell from the discussions presented in this book, is an idea that has drawn considerable interest in the context of the emergence of nanotechnology. Recently,

CCIT also has begun to define an initiative to look at international security and examine potential concerns regarding the military security issues in nanotechnology. This initiative would examine some proposed control regime models, considering both the possible benefits and the possible hazards of such control regimes.

CCIT feels initiatives will likely create international controls on the proliferation and use of nanotechnology in consequence of the weapons implications of the technology. CCIT is quite concerned that, in the course of creating such controls, the benefits of nanotechnology could be severely circumscribed or denied entirely. CCIT is also concerned that there could be serious infringements on the ability of persons and organizations to experiment in seeking new directions for society, or conversely in attempting to conserve existing values.

CCIT is also concerned that certain trends and developments might give rise to international conflicts. The approach that we favour in meeting the potential problems of a nanotechnology era is one in which a protocol for interaction between people is established. This protocol would maximize individual rights and liberties, while also attempting to establish the minimum effective types of control to prevent major harmful consequences to the world, which could (at their worst) be very serious indeed. However, in seeking to determine the nature of protocols that will serve these ends, the only ones that people will accept in the long term are fundamentally constitutional ones—that is, ones that balance whatever protection may be needed against people's liberties, in a predictable fashion with appropriate safeguards against abuse over time.

Politics of Technology

It is not possible to discuss the *politics of nanotechnology* today. As of today, there are no politics of nanotechnology. There is long-range speculation regarding a possible politics of nanotechnology. However, given the way politics works, there are not going to be any issues in nanotechnology until we get closer to an actual *nanotechnological assembler* or *proto-assembler*, or some kind of actual device.

The governmental means and ends and mechanisms that exist today to deal with technology will likely be the ones that will deal with nanotechnology when it emerges. Defense analysts have a term, "*come as you are party*," that refers to the fact that when war starts, one fights with the weapons that are in place at that time. The lead time to build and deploy a modern weapons system is considerable. Similarly, nanotechnology policy will be a "*come as you are party*," in that the *institutions*, *laws*, and agencies that will address nanotechnology policy and regulation, will be the ones that are in place before anyone takes it seriously—that is, the ones that are in place today or get created very soon.

A fundamental division between the forecasting function and the active functions is the funding of research, and the policy-and-regulatory-formation function. The former consists of the parts of the United States government chartered to perform long-range forecasting and to examine new technologies on the horizon, to consider what policy questions they might engender, how they might be addressed, and to make recommendations regarding appropriate courses of action.

Unfortunately or fortunately, depending on your point of view, these views and recommendations usually have only very limited impact on the actual emergence of the technology, or the actual policy measures adopted when those technologies emerge. In actuality, no matter what the long-range planning recommends, when the issue actually comes before Congress and the Administration (in this discussion, when a company or an institution creates an actual nanotechnology-related device), that is the point at which one must consider the issue to be raised *in vacuo*. None of the long-range forecasting previously performed will necessarily constrain the Administration or Congress regarding the assignment of regulatory responsibility or the nature of the applicable regulatory regime. Other, more short-term considerations addressed in this chapter have a much greater effect on such issues.

A Lesson from Space Policy

Some prior existing actions can bind, or at least circumscribe, the outcome of the policy and regulatory process. The commercial *space field* offers instructive parallels. Certain international treaties negotiated in the late 1950s and early 1960s (and finally signed and ratified in 1967), along with some implementing protocols added in 1972, did set a number of bounds on subsequent commercial space activity. Specifically, the Outer Space Treaty of 1967—a multilateral United Nations treaty between the United States, the former Soviet Union, and the United Kingdom, France, and other countries—contained a clause that effectively held that private activity in space was legal. In fact, President Kennedy delayed the original signing of the treaty in 1963 until the first commercially owned and operated satellite was launched, in order to ensure that a precedent for private property in outer space would be established before the treaty was signed. At that time, the former Soviet Union's position was that any private property in space should be forbidden in the treaty. The final text of the treaty did not accomplish this. However, as a compromise, a clause was inserted in the treaty, to assign unlimited liability for damage caused by objects in space to the "*sponsoring state*," a term that was defined imprecisely and permitted at least two differing interpretations. Subsequent legislation continued the ambiguity. As a result, when first began efforts to launch rockets on a private basis as entrepreneurs, the government's reaction was to say, in effect, "The Government has a direct stake in your activities. No matter what you do, the government has the liability for it, so we have to take a heavy role in your insurance requirements and all that drives them."

As a result of these prior treaty constraints, it was not possible to follow a maritime or aviation precedent in third-party liability issues in space commerce, although such a precedent might have permitted a more favourable outcome. The lesson for other emerging technologies is that it is possible to have international agreements negotiated well in advance of the advent of the technology. Because these agreements cannot accurately foresee the circumstances of the emergence of that technology, they can result in a binding of the hands of those attempting to create the regime at the time of emergence. In other words, bad foresight may be worse than no foresight. Currently existing international agreements (particularly regarding environmental issues, or in munitions control and *weapons proliferation*) may affect nanotechnology, and it is not even clear at this time which agreements those might be. When the technology emerges and we can actually ascertain its nature, and better know what the first incarnation of this technology will be, then we may discover that some of the basic laws of nanotechnology have already been written and agreed to. That will be more clear as we get closer to that time.

Players in the Government

The study or forecasting function is its own universe. Similarly, the active side divides into a number of universes. Among these, it is particularly useful to examine two sets of players: one are the sponsors of research, and the other are the policy and regulatory players. The research universe will be one of the first areas that impacts nanotechnology. The first issue within the research realm is allocations for basic and developmental research. (Given that Japan has begun significant funding for such activities, the question of why the United States government does not have such funding becomes an active, rather than a speculative question.) In the policy/regulatory realm, it will be the regulatory activities that, as commercial nanotechnology begins to evolve, will begin to affect those efforts.

In the research realm, the mainstream science research establishments (such as the *National Science Foundation* and the *National Institutes for Health*) allocate most of the civilian research funding. Neil Jacobstein's presentation on the federal research establishment constituted an excellent and detailed survey of this arena. Much basic research is done under the aegis of the Department of Defense. However, much of this research is not *applied research* (research that is actually developing or building weapons). Much is *basic research* because they realize that unless there is an understanding of the basic physical principles underlying technology, advanced weapons development must be limited. The Defense Advanced Research Projects Agency (DARPA), which is now the Advanced Research Projects Agency (ARPA), sponsors a wide range of research. Some of the service laboratories, particularly the Naval Research Laboratory, sponsor much basic research. There is also substantial intermediate development work performed at such laboratories, which is as applicable to commercial technologies as it is to military technologies. Laboratories sponsoring such intermediate research could be early participants in nanotechnology research. Some important precursor research relevant to nanotechnology is currently being sponsored at the Naval Research Laboratory.

A beneficial aspect of the defense research arms is that a substantial amount of the basic and early stage applied research is done in a nonclassified environment. Over the last decade, the Strategic Defense Initiative Organization, or SDIO (the so called "*Star Wars*" organization, which is now the Ballistic Missile Defense Organization (BMDO)) has sponsored interesting basic and applied research, much of which has been good research. An interesting aspect has been that SDIO has been more open to different and unusual proposals for research than many of the established research institutions, as they relied more on the individual discretion of the granting officers. This has been especially true in their "Innovative Science and Technology (IS&T) Program," which has been a source of support for many innovative high-technology entrepreneurs over the last decade. Much of this research has very little to do with the ultimate "*Star Wars*" system, but rather are interesting pieces of technology that some officer decided would be a good thing to do.

The White House offices impact both basic research and long-range planning and thinking. The Office of Science and Technology

Policy (OSTP) three decades ago was the President's science advisor. Today, the Director of OSTP is called that, but originally he literally was that (that is, a scientist whom the President would occasionally call to ask for advice). In the way of Washington, they presumed, "Well, you have to have a secretary." So, they got him a secretary, and then, "Well, he ought to have an assistant." Before you knew it, there were 30 people in an office. It was enshrined in the Old Executive Office Building (next to the White House), and then they had to spill over to the New Executive Office Building. Soon it was bigger than cabinet departments were in the day of Thomas Jefferson. The bulk of what they do is forecasting and thinking, and they issue papers, and it does affect decisions.

The Office of Management and Budget (OMB) has, in reality, more impact because they have the final say on the Administration's decision regarding what gets funded. The Program Area Director for Science, Technology and Space, inside the OMB, has much more to say about what the Administration will and will not do in space and science than OSTP. People often overlook that fact, but it is well-known. It is the "*Golden Rule*": the one who controls the gold makes the rules. This situation is no exception.

In considering the Congressional side, one must remember that under the American system of government, the President proposes, but the Congress disposes. Congress makes the final disposition of funds. There are programs that agencies never wanted, but that Congress tells them they are going to have any way. The Osprey V-22 vertical take-off tilt-wing aircraft, which the Pentagon never wanted, is an example. The decisions in Congress are, in essence, made by a few key Congressmen and their key staff members, especially the permanent committee staff that report to the chairman, as opposed to the staff of an individual congressman.

The authorization committees will say, "We need a program in X, Y & Z. Let us allocate $100 million to this program this year, and we will allocate the money to be spent in District A and District B." To accomplish that, the authorization committee generates an authorization bill that authorizes that money. However, that money is not real until the appropriations committee then says, "Yes, we will spend the money," or, "No, we will not spend the money." Their effect, on the Congressional side, is somewhat equivalent to that of the OMB in the Executive Branch process. Of course, the full House and Senate must approve the decisions of both authorization and appropriations committees (just as the President must approve the decision of the OMB), but a failure on either part to make such approvals is a rare and remarkable event.

The Office of Technology Assessment (OTA) is an independent office reporting to Congress (as opposed to the Executive Branch), which performs long- and short-range studies. Their charter is to examine any particular technology question and draw up scenarios that attempt to foresee consequences of alternative courses of action. In effect, they generate stories that say, "Well, if you do X, this will happen. If you do Y, that will happen." They are not sup posed to make recommendations, although often you have a report that, in essence, says, "If you do X, it will bankrupt the nation. If you do Y, you will blow up the world. If you do Z, everything will work out great, but we are not making any recommendations. Go do what you want."

Analyzed operationally, one might be tempted to say that the primary function of the OTA is to provide jobs for technology-oriented staffers of the party not holding the Presidency, since they cannot get any jobs in the Executive Branch when their party is out of power. Following this, one might predict that you will now see a migration of Democrats from the staff of OTA over to actual operating agencies.

All of these elements affect the outcome in varying degrees. The locus of early research may determine some elements of the ensuing governmental reaction. For example, if a substantial portion of basic nanotechnology research is funded by NIH, that research will be oriented toward medical and health applications. If it is funded through NSF, it will be more generic. If it is funded through the National Institute of Standards and Technology (which is part of the Department of Commerce), then it will be probably more commercial- and manufacturing-oriented. If it is funded through DARPA, it may very well be oriented toward defense applications. It is quite possible that there will be parallel programs running in all those places, as well as independent research financed in the commercial sector.

Looking at the executive departments themselves—those that are headed by cabinet Secretaries or Administrators (as is NASA, which is an independent agency)—a number of the different cabinet departments have substantial interests in technology. The Department of Commerce has the National Institute for Standards and Technology, formerly the National Bureau of Standards; the Patent Office, which has its own independent and very strong impact on the way technology develops; the Bureau of Export, which determines what can be exported and what can not be legally exported (aside from the weapons-export issues addressed by the State Department rather than by Commerce); and the Department of Energy which runs National Laboratories such as Livermore and Los Alamos.

The *Department of Energy* has the Federal Energy Regulatory Commission (FERC), which regulates many energy-related activities. If nanotechnology first emerges as part of an application in an energy production area, FERC may be the agency that does the first regulations on it.

The *Office of Munitions Control* (OMC), part of the Department of State, administers the International Traffic in Arms Regulations, which control many technology enterprises not normally thought of as weapons (such as commercial communications satellite technology). OMC is one of the leading candidates for being the controlling agency for nanotechnology. It must be understood that "export" in this sense is not limited to taking a device, putting it in a crate, and shipping it overseas. It can mean putting the plans or information relating to a device or technology in an envelope and sending it overseas. Or, it can mean taking information that is critical to developing a technology and publishing it in the United States, since once it is published here, it is available around the world. OMC, by this logic, has an enormous amount of power. Some speculation has been published in the legal press that the OMC's charter and the Munitions Control Act of 1950 may, in fact, be unconstitutional in that they violate the First Amendment. That point has never been put to a firm test. It may be soon.

NASA is an interesting anomaly because work that would ordinarily be performed under the aegis of NSF (if it is space-related) will often be performed by NASA.

The *Environmental Protection Agency* (EPA) has an enormous amount of regulatory power and, since the release of nano mechanisms in the environment also potentially falls under the jurisdiction of the National Environmental Policy Act, EPA could end up being the regulatory agency.

Since the *Federal Drug Administration* (FDA) has jurisdiction over medicine and medical devices, if the first applications of nanotechnology are in medicine or medical devices, they may be the lead regulators. As John Doerr has said in his presentation, it takes about a $100 million to get a drug approval through FDA. If a nanotechnology device is interpreted, as it may be, to be a drug rather than a machine, then this could mean that the product will not be seen first in the United States, but rather elsewhere.

To emphasize the general implications of the previous point, when the first *nano device*, *nanomachine*, or *proto-assembler* device comes on the market, the regime that is going to control it is probably already in existence. It is not clear which one it is going to be, but the three top candidates at this time would be the EPA, OMC, or the FDA. None of those is the most suitable, in my opinion, but quite often, from the standpoint of businessmen, if the regime is already in existence and already understood, that is a plus. You can go to a lawyer and get an estimate of how much it is going to cost you, and how long it is going to take to get an approval through. It may be that if a company has a nanotechnology device that is close to market and has (let us say for the purposes of discussion) a medical application and a manufacturing application, the company may (out of caution) make a decision to bring out the manufacturing application first because the regulatory path is easier. Such considerations will be real, and they will arise.

Now that Al Gore, who held hearings on nanotechnology in the summer of 1992, is Vice President and head of several important councils, it is possible that nanotechnology will begin to emerge as an item of discussion. One of the actions that has been discussed was that of having the OTA include nanotechnology in its study areas. With Vice President Gore's party in power, this task may shift from OTA over to one of the Executive Branch agencies.

The government will not do much beyond forecasting and perhaps sponsoring of long-range research until an actual company with an actual project (or alternately, an actual federal agency with an actual internal project) is about to do something. The agency with the most proximate charter will have the strongest incentive to act. There may be a little "*gold rush*" of agencies at the time when nanotechnology approaches implementation, or maybe a big frenetic rush, depending on how important it is seen as being.

Certainly, when commercial space emerged as an issue in 1981 and 1982, there were prospectors from the Department of Commerce, the Department of Transportation, NASA, FAA, Bureau of Alcohol, Tobacco and Firearms (in its capacity as the regulator of explosives), and about 23 other agencies (including Department of Agriculture and Department of Interior), all investigating the possibility of asserting jurisdiction and preparing arguments to support their cases.

Influencing the Government

In influencing the governmental reaction to *nanotechnology*, CCIT would like to do a number of things. When the "*gold rush*" starts, we can have a considerable impact by evaluating the various options for regulation (or approaches to regulation that are presented by the situation at hand) and trying to work with the agencies we think are most appropriate. This is a classic tactic in such situations. It may be possible to propose that a new agency be established within one of the existing cabinet Departments and to promote the idea of that agency being created and given the jurisdiction.

That course of action has certain benefits, because in doing so, it creates a little more leeway in suggesting what the agency's regime is going to be like. A party active in the field has a lot of leeway in suggesting what the regulating regime is like, no matter who gains jurisdiction, because typically you have some bureaucrats who have just been given a new role. They want to make the industry happy. They do not need the industry going up before Congressional committees every six months and saying that this decision is a disaster and that the agency in question ought to have their jurisdiction taken away. The agency does not desire this. They need the industry to say, "These people are doing a great job and their budget should be increased." So, they will listen to industry. It is a "*mutual capture*" process.

The agency will let the actors in the field write the regulations to some extent. If you pay to have the legal work done, to have a draft regulation written and present it for the agency's edification, the ideas and approaches presented will often get substantially incorporated in the draft regulations that are created. Everyone likes to have someone else do work for them, and bureaucrats are no exception.

This is not necessarily a *pernicious process*. In reality, many people in the federal government realize that people out in industry (or out in the front lines of the research) know much more about their work than they themselves do, and they listen to suggestions. From the public interest perspective, the problem with this process of mutual capture is that it cannot prevent the cooperation from including cooperation at the expense of potential new entrants to the field, and thus at the expense of the public.

CCIT is planning to pursue multiple, complementary courses of action. For the present, CCIT will be active in the forecasting world, CCIT is seeking to influence this realm by advancing its scenarios. Because of CCIT's status as member of the Foresight family, and, thereby, its access to the best, most cutting-edge knowledge on nanotechnology, CCIT is able to have the most accurate understanding of the prospects for this area, and, therefore, the most useful scenarios. As the research advances, CCIT will foster discussion of what kinds of research are done, what kinds of institutions are funded, and by what means funding is allocated. Subsequently, when actual regulatory questions arise, CCIT intends to be present with suggestions on what to do.

Industrial Policy

To give an example of what technology policy debate is like, we want to make a few comments about a topic which four or five of the other presenters here have mentioned a lot, but about which there has not been much substantive discussion. That lack is not the fault of the presenters, because this issue has been discussed mostly in a circular fashion for the past fifteen or twenty years in Washington. This is the so-called "*industrial policy*" issue.

To begin with, it is quite a bit of a task to define "*industrial policy*" precisely. At its simplest level, industrial policy is the set of governmental policies that affect industry, including the kinds of issues that have just discussed—funding levels, mechanisms, and categories for basic and applied research, regulatory regimes, and similar issues, such as been advocating that the government do, or in some cases, not do.

However, more recently, another, more specific understanding has come to the fore in the debate—industrial policy in the sense of an integrated government plan to foster desirable industries and technologies. Many people have advocated industrial policy in that sense recently, especially since the Clinton Administration (which has promised to have an industrial policy of that nature) has taken office. There is an attitude that says, in affect, "Now that we have an industrial policy, we have made it. Our problems are over." In fact, once it is determined that the nation will have an industrial policy, in many respects your troubles have just begun. Because you must define exactly what kind of industrial policy you have, you must define how you are going to run it and which course among many you will take. It is important to keep in mind that the success or failure of many, if not most, technology companies or initiatives will now hang on the nature of these decisions.

Advocates of industrial policy often point to foreign successes that they attribute to successful industrial policy. However, industrial policy has had a very mixed track record overseas. How many of you use a computer of any sort in your work? Please raise your hands. (everyone in the audience raises hands.) How many of you use a Machines Bull mainframe as your computer? (hands are raised.) Nobody? How many of you even know what "*Machines Bull*" is? (hand is raised.) Machines Bull is the European "*national champion*" computer company. This was the product of pan-European industrial policy, and particularly French industrial policy. The European governments, and particularly the French government, poured billions of dollars of French and European resources into trying to make Machines Bull an equivalent competitor to IBM. For various reasons, this effort was, in essence, a complete failure. The French government still uses many of these machines, and, to some extent, European businesses do, partly because they get pressured by the government to do so. One must ask, "What is the opportunity cost of putting all that money into a massive failed project? What could have happened if that money had been left to industries to spend for themselves, or spent by the government in a different fashion?" Industrial policy is not an automatic success. It has had quite a mixed track record overseas.

One must note, and wonder about the Clinton Administration's promise of both strong industrial policy and strong actions on creations of jobs. They have had a heavy emphasis on job creation. Countries with strong industrial policies often are very poor in job creation. The continental European states are considered one of the classic examples of strong industrial policy. They have higher unemployment rates (particularly for young entrants in the job market) and they have had a much poorer record on job creation over the last 15 years than the United States. This includes France, where they artificially maintain a high percentage of people in the agricultural sector relative to other industrial nations, thus absorbing a lot of people that would otherwise be in the industrial job market.

Japan has a structure that absorbs a significant number of people, not through deliberate industrial policy, but through custom (which maintains an inefficient retail apparatus), as well as a highly subsidized agricultural sector. The result is that many Japanese are underemployed at low wages. Those people are being held off the industrial job market, plus, because of Japanese social customs, most women are kept out of the job markets. It is not clear if they had not maintained those social differences that they would be much better than the United States in terms of unemployment. It is not clear that in cases where countries have industrial policies and strong successful competitive industries, whether the successes are because of the industrial policy or other factors. It is the old question where rain and wet streets are associated, but you cannot make it rain by wetting the streets. So, you must be careful in attributing cause and effect in examining industrial policy.

South Korea has a strong industrial policy with central planning. They have had highly successful industries in automobile, shipbuilding, and a number of other areas. Taiwan and Singapore have had very different types of industrial policies, but they have also been very successful in job creation, competitiveness, and good foreign trade balance. One might argue that these are examples of success through strong industry-government cooperation. These countries have gone from being at the income level of Haiti in 1960, to having giant foreign trade reserves and now pressing into the ranks of developed nations. So, clearly there is a correlation—or is there? Consider that South Korea, Taiwan, and Singapore are three of the so-called "four tigers." The fourth "*tiger*" is Hong Kong, which has had an equivalent (or greater) rise of success in competitiveness, but has had no industrial policy whatsoever. So, it is not entirely clear that the successes are because of the industrial policy practiced by those governments.

Another factor to consider is the legal and cultural systems of those countries. It is worth noting that the countries with legal and cultural systems most similar to the United States that have tried to implement industrial polices have had the least successful outcomes from it.

Canada probably has the political, legal, and cultural system closest to the United States, and it has had a very dismal record in industrial policy. The United Kingdom has had a strong industrial policy for 30 or 40 years, and, in addition, has poured a considerable amount of money into science and technology education. They still have excellent universities, excellent education programs, and they have been totally unable to turn this advantage into competitive technology or products. One must ask, "If we are to have an industrial policy in the United States, what aspect of our law or culture might we have to change to make this more like countries in which it has seemingly been more successful?"

Our liability *law system* creates many hazards and traps for companies putting forward new technology. It is not clear that we could have a successful industrial policy unless our liability policies were more like that of Japan, Germany, and France (that is, tort reform). However, will we have reform in that direction? Consider that trial lawyers, who have formed a well-funded lobby against tort reform, were the largest early donors to the Clinton campaign. This fact suggests that tort reform of the sort needed to support industrial policy, is unlikely for the current presidential term.

Antitrust is another area. We have an *antitrust system* that is unlike that of any other country in the world with strong industrial policy, and it has in the past worked against the kind of industry-to-industry cooperation (much less industry-to-government cooperation) that is typical of industrial policy systems. Are we going to change the antitrust system? This is not evident. Historically, the Democrats have been the pro-antitrust legislation party.

Continuity in government is another area. Typically when Europeans are promoting an industrial trade product, their trade officers and their project officers remain the same despite changes in party or administration. Britain or France typically change about 300 civil servants at the time of the trans'tion from one party to another, and their parties are more extremely polarized than those in the United States. The United States Government typically changes about 3,000 officials. We have had space trade negotiations with the European Community, China, and the former Soviet Union, which are all extremely important to the success of the United States launch industry. Do you know how many people in the United States government have been with those negotiations since the beginning, and have a full, continuous understanding of the issues? One—exactly one man, and unfortunately for the prospects of continuity in this area, he is a Republican, so he is unlikely to retain his position. He is very likely to be replaced. If we have an Administration that advocates continuity in industrial and trade policy, should they not be keeping him? Should they not be keeping the person with the long track record?

Another example can be seen in the negotiations over the *Airbus airliner*, where the United States has been negotiating with the European Community since the late 1970s. The key person at the Aerospace Industries Association who had the responsibility for tracking those negotiations. She observed, "The Europeans have had the same team from the beginning up until today. The United States has switched teams every time they switch an Administration, and quite a bit in between." So, are we going to develop the tradition of continuity in government necessary to make this continuity possible? It remains to be seen.

"*Research monocropping*" is a problem. Just as a *monoculture* in forests or other agricultural areas is bad for the ecology in the long-term, which can be defined as "a tendency to concentrate all your research dollars on a common set of programs, a common set of directions." This is very bad for research. Monocropping in research means that you do not try the different veins of ore. You do not try the different paths. However, the incentives in government funding are toward research monocropping, because there is a prevailing assumption that the government plan is going to be able to identify all the productive areas of research, and fund them all. The ones that do not get included in the master plan, by definition, do not get the resources.

Right now we have considerable diversity because the research activities in different parts of government are quite uncoordinated. The beneficial effect of this is that it permits different directions in research. NSF allocates its research funding by peer review. DARPA largely leaves it up to the discretion of the program managers in the areas, so it goes by individual judgment. SDIO IS&T basically does the same thing—they give even more leeway. If there were a "*Rationalized Plan*," there would be a danger that the different approaches funded for very different reasons might all be brought into a single coordinated plan, which would mean there would be less room for different ideas.

Another problem is leader-versus-follower strategies in industrial policy. Industrial policy has historically been better for followers than for leaders in technology. The South Korean government said, "We want a steel industry." It is relatively easy to determine what goes into a steel industry. One needs steel plants, limestone, this and that, but it is all well-known. One can copy an existing foreign industry, incorporate the best of known current technology into it, and know how to put that together. One cannot say, "We want a nanotechnology industry," and proceed to put one together, because one cannot know yet what is going to be necessary for a nanotechnology industry. There is no model to copy. The states that have been most successful in industrial policy have generally been states that were catching up, and that were going into an existing industry and following the lead of other nations. There have been very few examples of nations that have gone out and seized leadership in a pioneering technology industry through industrial policy.

It is worth turning to Japan briefly because many people have an idea of Japan as a country with a successful industrial policy, which is a bit reminiscent of the caricatures of Japan that occurred during *World War II*. Because Japan enjoyed all these tremendous military successes for month after month during the first two years of the war, many people assumed that there was a Fu Manchu-type central genius in Japan who was planning the whole war, through which everything was tightly coordinated, and which provided that the Navy would strike here and the Army would strike there. After the war, when we were able to read the actual plans and interview the officers, and after participants published their memoirs, it became apparent that the Army and the Navy were virtually independent of each other and of any kind of central government coordination. Both were proceeding adventitiously. Admiral Yamamoto, perhaps the most perceptive officer on the Japanese side, made a famous statement at the beginning of the war. He said, "I can run amok for two years. After that, you had better negotiate." In reality, there was no plan. Japanese forces probed adventitiously in every direction. When they saw weakness in a target, they pushed, and it fell.

To some extent, Japanese industrial policy successes are like that. *The Enigma of Japanese Power*, as a general discussion of Japanese social and political institutions relevant to the industrial policy debate. A number of different, strong institutions exist in Japan, and strong coordinated sectors of government exist in industry, but they are not universal. There is no Japanese master plan. Japanese companies, and secondarily governmental institutions, are pouring money into a diversity of efforts and directions. Some of these have been very successful, and some of them have been great failures. To imagine that MITI is this same kind of omniscient central planner who knows all, sees all, and plans all, and is responsible for all these victories, is just as fallacious as it was during World War II. The world industrial market, at the level addressed by Japanese companies, is too complex to be dealt with by a central plan.

What many industrial policy advocates have been presenting is a kind of cartoon caricature of Japanese industrial policy, which is what some advocates would like to see implemented in the United States. It does not necessarily have that much correspondence to what is going on in a country like Japan. In a lot of ways, South Korea has a program that is like what people imagine Japan's to be, but South Korea deals much more with catching up on old technologies than with new technologies, and has historically chosen to compete in many fewer sectors than Japan.

In closing, it is also worth raising the issue of the Iron Triangle problem—the problem of political allocation of funds as affected by the interlocking interests of local beneficiaries of government funding, their representatives in legislative institutions, and the Executive Branch persons who depend on that legislature for funding. When new technologies are emerging in many new areas, one must invest in multiple paths. One must be prepared to switch support from area to area rapidly, because one can be pursuing a research path, and continuing on that path, and it will appear very promising. One will be mining the vein, to use Howard Landman's metaphor, and it appears to be fruitful, but then suddenly it dwindles, while an expected little side vein actually proves to be the route to the mother lode. In such cases, one must be able to switch one's resources from the dry hole to the promising vein.

When you have a central plan, this is hampered, as researchers become invested in a particular approach, and people began to obtain benefits, in effect, by mining the dry hole. The money will go to the research institute that is taking a certain approach and happens to be located in the Congressional district of the chairman of the committee. If that research area is becoming unpromising, one may need to reprogram resources to a different institution and a place that has the experts in a very different, desired area. Consider, then, the situation of the master planner who must say, "We must adjust the master plan. We must reduce the funding in Mr. Chairman's district, and we must reprogram it over to the district of Mr. Nobody." To say the least, very strong incentives exist against that happening.

The United States, government could not spend any money productively on technology. The government can do intelligent things in the way it spends money, especially by being a smart customer in buying those things the government actually needs for its own purposes, which can do a lot to stimulate useful technology and technology development in industries in the right areas. However, we should think very carefully about spending such money. My purpose has been to raise some points and, in doing so, to make clear that industrial policy is not a magic wand, or a magic solution to all problems. Before one could expect to create a useful industrial policy, many real problems must be discussed. CCIT intends, among its other tasks, to promote discussion on such issues.

12

Virtual Molecular Reality

The great thing about being a professor at MIT is that you get to have the best teachers, like Eric Drexler. They are called "students," and if you pay attention to them very carefully, you can learn a great deal and even get credit for some of their work. Once, Tinkertoy built a structure of Tinkertoy rods that reached to the top of a hotel lobby. The adults tried to figure out how a person so small could build something so big. My problem was to figure out how people so large could think so small. If you remember Tinkertoys, with its wooden rods and 10-holed spools, you may be showing your age. They still exist, but, nowadays, LEGO is more popular. In recent years, LEGO has added hinges and other new parts so that you can join things with more diverse angles, but this makes the system less elegant.

The important thing about Tinkertoy—and other such construction sets—is not merely that they help you to understand structures, space, and geometry. They also convey a profound idea that could help children to understand certain other important domains. Such toys teach you the concept of a "*Universal Set of Operations*," the idea that a limited set of primitive operations can generate an infinite and rich variety of composite structures. For example, this is why chemistry is so rich, although it, too, like Tinkertoy, starts out with nothing more than a few kinds of atoms and a few kinds of bonds. As another example, this is why mathematics is so rich, although it starts out with nothing more than a few axioms and a few rules of inference. And this is same reason why Drexler's approach to *eutactic nanotechnology* is so rich.

There are some, who would object that these elegant toys might create for the child a sterilized reality. "*Better*," those critics might maintain, "to give the child more realistic, less ideal materials. Give them things like clay and paint, things you can squeeze and smear, so their imaginations will be less constrained." There is something to be built with mud and sand. However, that with those "*more natural*" materials, one can never go very far beyond certain levels of complexity.

The parts made from wood, saw, and nails never fit quite properly, and the errors would accumulate. Thus, when tried to build with wood, each piece would end up out-of-size by at least a sixteenth of an inch, because of my forgetting the width of the saw. It was only in much later years, when shaping steel with milling machines, that this kind of error quite disappeared. Instead, the piece I was working on would end up at precisely the desired dimension to within one ten-thousandth of an inch—but sometimes with an error of precisely plus or minus one inch! This is because, as machinists know, the most common mistake is to miscount the turns of the coarse-feed crank.

A universe in which things worked exactly as they were supposed to work. When you build things with Tinkertoy, Erector, or Meccano sets, a proper conception yields things that perform—and if they don't, then your scheme was wrong. You should not need to file or shim or bend those parts to make them fit—the way that everything else is done. Two objects assembled with those kinds of parts are exactly the same, or clearly not—precisely as it is with molecules, mathematics, or programming.

You know how, in the "*real*" world, things scarcely ever work very well? It does not help when adults explain, "It is not important to reach your goal. It is only important to do your best. It is the journey that matters, and not where you get." That is not wisdom, in my view, but only the sour-grape excuse of broken-spirited mortals. It is only partly because of nanotechnology's elegant new ideas and possibilities.

Quantum Certainty

In *espousing nanotechnology*, both Eric and Tinkertoy have had problems with the academic establishment. Many otherwise excellent colleagues have objected that nanotechnology might be impractical because of quantum uncertainty. "You cannot be sure where the atoms are," they would complain, grumbling about the uncertainty principle, tunneling, and strange correlations. In *Nanosystems*, Drexler treated this carefully by calculating many such uncertainties, and showed that those skeptics were usually wrong. At ordinary temperatures, the thermal effects tend to dominate the quantum effects—and tend not to be very serious. In the rod-logic structures that he designed (without any redundancy), the most serious problem is damage from cosmic rays—and that would happen classically, too, if exposed to such energetic particles. Chemical structures like DNA can be stable for many millions of years, if there are no other destructive processes.

In any case, these objections are unconsciously based on a peculiar bit of philosophy that has insinuated itself into our popular intellectual culture. We see it in statements such as the following:

"In the world of classical mechanics, everything worked like clockwork, with deterministic certainty. Then quantum theory changed our view, so that now we understand that all is uncertain and indeterminate."

Almost everyone accepts this view of history. However, now I will explain why one should take an opposite view. Uncertainty was inherent in the classical view, while only quantum theory showed that things could be depended upon.

Let us explain this seeming paradox. Before quantum mechanics, the dominant idea was that matter was made of particles that interacted through inverse square laws, gravitational and electrical. Consider a solar system, for example, as described by Newtonian physics, that has a heavy object in the middle, and several lighter objects surrounding it. Such small systems appear to be rather badly behaved—and in the particular case of the solar system in which we live, no one has been able to show that it is stable. For example, Gerry Sussman (another "student" of mine) and Jack Wisdom at MIT seem to have shown that the orbit of Pluto is chaotic, and, so far as it is known now, *Pluto* may eventually get thrown out of the solar system. We Earth-people might not consider this a serious loss. However, consider that our largest planet, Jupiter, has enough angular momentum that, given suitable coupling, it could hurl Earth itself into outer space.

Thus, solar systems are unstable things. The same very likely would be true for atoms and molecules if they were similar, if their nuclei and electrons interacted by those classical laws. Even if each such atom by itself happened to be stable, when any two such atoms approached one another, the electron orbits inside them would soon be perturbed, and one or both atoms would soon break up. In the world of quantum theory, however, each atom is stable—completely unchanged—until there occurs a transition jump. The result is that we can have molecules with covalent bonds, in which the electrons maintain precisely at specific energy levels for billions of years.

Thus, contrary to what our science teachers tell our kids, it was in that old Newtonian World that almost everything would be unstable and indeterminate, whereas it is quantum mechanics that makes possible chemistry, life, and nanotechnology. It is because of quantum states that you can remember what you had for breakfast. This is because the new neural connections made in your brain can persist throughout your day. When something more important occurs, you will remember it as long as you live. Everything that we can depend upon exists because of—it needs a name—because of "*Quantum Certainty.*"

So, the next time you hear those philosophers (or retired, burned-out physicists) speaking about uncertainty, ask them if they realize that only Quantum Certainty makes anything we know persist. Is there any way to do the same within the so-called deterministic world of classical physics?

Missteps toward Micromachinery

Tom Etter and Rollo Silver discussed about the idea of making a tiny machine shop, with miniature lathes and milling machines. The idea was to use these tools to build mini manipulators that could work on a ten-times-smaller scale. (We all were partly inspired by Robert Heinlein's prophetic 1950 novel *Waldo*, whose inventor hero had a disease that made his muscles extremely weak. Fortunately, he was wealthy enough to have a satellite built so that he could live in zero gravity. Because he felt so isolated there, he invented teleoperators so that he could do what he wanted back down on Earth.)

Anyway, once you had made those smaller manipulators, you would use them to build another machine shop yet 10 times smaller, and so forth. Each such step might need some new technology (for example, because the previous generation of bearings and motors might not work well at the next smaller scale). We assumed that each such set of problems might be overcome in a decade of work. Now observe that a nanometer is only 10^{-9} meters—that is, a mere 10 of those decade-steps away. If you started today, you could easily assemble, with off-the- shelf parts, a 1-cubic-meter machine shop. So, if all went well, in just 80 years you would have an atomic-scale assembler. We have not quite made that much progress in the 40-odd years that have passed since then, but we might be able to make up that time by discarding all those intermediate steps and going straight to eutactic technology. No one yet has demonstrated a foolproof way to build the first such assembler, but that is surely only a matter of time.

Early in the 1950s, Etter and Silver also conceived of "computing cloth"—an idea inspired by neon bulbs. Their scheme was to use closely spaced electrodes to move tiny plasma discharges inside a rare-gas atmosphere. (It requires a high volt age to start an arc, but lower voltages can make the arc move from one location to another, and they showed how to make ANDs, ORs, and Inverters this way.) The goal was to weave all the circuits into a microscopic insulated-wire cloth, while creating all the logic gates simply by removing the insulation at appropriate points. Etter and the Silver equipped the lab with vacuum equipment and glassblowing stuff, and eventually built a non-so-small shift register that used these gates. This was long before the earliest integrated circuits. Unfortunately they could not raise enough capital, so the project was abandoned.

Not long after that, Silver continued to dream about how to economically "*stamp out*" computers. He actually built some small hydraulic computer elements, by making millimeter-wide grooves and holes in multiple layers of plastic sheets, and placing little rods or balls in some of those grooves. When the assembly was pressed together and connected to a water supply, it became a computer, with circuits that worked at 30 Hz, powered by a 3-inch high column of water. Again, no one seemed interested in his little "*hydro flip computers*," although other forms of hydraulic logic were becoming popular for designing radiation-immune machinery.

Also, in 1955, with help from Ned Feder (more recently of science-fraud fame), Meccano developed an elegant electrically controlled micromanipulator with which you could easily write your name in submicron-sized letters. It even had a way to estimate the force at its tip by vibrating the probe and viewing the amount of deflection. It simply never occurred to me to try make it sensitive enough to resolve individual atoms.

What to do Next?

Drexler's schemes for making machines was the use nothing but latches and rods—like making computers of Tinkertoy. Indeed, Danny Hillis and a group of other students made one of those once. This is stuff that clearly could work, and that even a child could understand. If at any stage you encounter a bug, you could always scale the whole thing up. People pointed out that Drexler could make faster computers by using electrical circuits instead of rod logic, but this would surely lead to serious new problems. Strongly bound atoms do not jump around much (though we better be careful of hydrogen), whereas electrons might tunnel all over the place. If we decide to take that route, we must select some functional units to mass-produce.

The alternative to assembling parts is to develop a more integrated technology, through which a system is designed, simulated, and then fabricated in a non-modular fashion. In the early days of "small-scale integration," computers were made by mounting separate logic modules onto a large, planar "*mother board.*" The advantage of this was to make it easy to replace failed components. An amusing aspect of this was that by the mid-1970s, the price of small-scale logic chips had dropped down to 15 cents or so, but the sockets to hold them cost 5 times more—and often the connectors between the boards ended up costing more than the components themselves. When we begin to build our first nano machines, will we begin with small modules to be assembled later, or should we try to do everything at once, by making the whole machine in one piece?

Skeptics

It was fascinating to see how skeptical people were, over the past decade, about whether nano machines could be possible. This era is close to its end, although there will always be those who make their living (or at least their reputations) by being skeptical. In the field of *artificial intelligence* (AI), there are always one or two such critics, but usually not many more. Perhaps this is because there is room for only a few best-selling books in this area. In any case, it seems quite strange for anyone to argue that you cannot build powerful (but microscopic) machinery—considering that our very own cells prove that such machines can indeed exist. And then, if you look inside those cells, you will find smaller machines that cause disease. Most arguments against nanotechnology are arguments against life itself.

When skeptics assert that no matter what else we make our machines do, there is no way we could ever make them be conscious, or sentient, or anything like that. However, when you ask those critics precisely why this should be a problem—that is, what aspect of conscious ness is beyond a machine—the answers that come back are almost identical with the arguments, a century ago, against molecular biology. "No mere arrangement of chemicals could possibly become alive." "*Why not*?" "Because you cannot reduce the vital spirit to anything so understandable." When this exchange is translated into the domain of AI, the phrase "vital spirit" is replaced by the term "*consciousness.*" When you ask for more details, the discussion becomes exceedingly vague, and the skeptics cannot seem to tell you what they mean—except that machines can not be "*self-aware.*"

So, then, try asking these skeptics to what extent, and in what senses, are they aware of their own selves. First they will answer, "Clearly, I know that I am here." They could say that just as well, about the presence of their shoes. So, now ask, "Please tell me what is happening inside your own mind? How did you choose the words you just said? How did you put them in sentences? How did your brain recognize? How did it know what meant by them? And tell me, how did your brain decide which shoe to put on first today?" No human can answer any of these, because we simply have no access to how we choose one thought or expression over another, or why we first put on the left shoe. Whatever consciousness might be, it simply does not do the things that all those skeptics claim it does. So, it seems to me that most of those critics go wrong at a truly critical point. If you think of consciousness as being aware of the processes that underlie your ability to think, then there is scarcely any reason to suppose that people are conscious at all. (If they were, we would not need psychologists.) Now, contrast this with an AI program running under a LISP interpreter that has its functional "trace" feature turned on. Then, whatever that machine might happen to do, you can always interrupt it by asking, "*Why did you do that*?" The machine is capable, in principle, of returning descriptions of how each decision was made, at least until it runs completely out of short-term (program-stack) memory.

My guess is that the most significant difference between the ordinary chimpanzees and the dressed-up chimpanzees is that, at various times in the past few million years ago, our brains augmented their capacity for "*short-term pushdown*" memory. That made it possible for certain parts of the brain to "*replay*" descriptions of what happened in other parts of the brain, and these enable us better to keep track of things, to analyze which thought processes had been most effective, and to remember methods that went wrong and still have some chance to straighten them out. It was the same, last century, when "self-reproducing" seemed obviously beyond what any machine could do. Today, we know dozens of ways, and all of them seem trivial.

Gauging Progress

One kind of complaint about AI does have a little more merit. Many people assert that in the 1960s, AI seemed to be advancing rapidly, but it no longer is keeping its promises. The work appears to have slowed down so much that there has to be something fatally wrong. There are several answers to this.

The simplest answer is that we solved many easy problems in the first few years. Now we face harder problems, so naturally progress will tend to be slower. Certainly this is partly true, but not nearly as much as those critics perceive. This is because, in every field, progress in the past 10 years rarely seems so fast as before, simply because you cannot yet tell which recent work was important. Certainly, during the 1980s, many of those who considered nanotechnology to be a science-fiction pipe-dream were wholly and simply unaware of progress already underway in such domains as building artificial catalysts, and improving atomic-force scanning microscopy.

13

TRANSMISSION OF BIOMOLECULES

The cells is a very complex and its function depends upon the integration of the activities of many genes. The genes may interact within the organism. The genes control various traits is an organism through a control exercised on the developmental processes. Such a control is due to the synthesis of enzymes or proteins. It is a well-known fact that the synthesis of protein is under the direct control of gene or genes. Thus we see that the genes express themselves through the synthesis of enzymes and it was demonstrated for the first time 1941 by *G.W. Beadle* and *E.L. Tatum* while working on the biochemical mutations is Neurospora. Based on their this work, *Beadle* and *Tatum* proposed a concept called "*one-gene-one-enzyme*" hypothesis. This hypothesis means that all the steps which transform a precursor substance to its end product that is ultimately expressed into a structural and functional phenotypic trait, constitutes a *biosynthetic* or *metabolic* pathway, and each step of a biosynthetic pathway is catalyzed by a specific enzyme which in turn is synthesized under the control of a specific gene.

Biochemical genetics has been widely studied in haploid organisms whose growth requirement are known. Any error in the metabolic pathway is expressed in the form of disease as in human beings.

GENETIC CONTROL OF METABOLISM IN HUMANS

Early evidence for the relationship between genes and enzymes came from the work of the physician *Archibald Garrod* in 1909, the results of which were published as a book called *Inborn Errors* of Metabolism. Garrad was interested in human diseases that had an apparent genetic basis. He postulated and rightly so, that biosynthesis and degradation of a product takes place in several sequential steps and these steps are controlled by genes, one gene controlling one step. He suggested further that any of these genes mutates in the metabolic path, all the other reactions ahead of it, stop and there is accumulation of the substance which the intact gene preceding the mutant one, synthesizes as shown under the following subheads:

Phenylketinuria or Folling's Disease (PKU)

This disease is caused sue to the absence of enzyme which catalyses the conversion of phenylpyruvic acid into hydroxy-phenylpyuvic

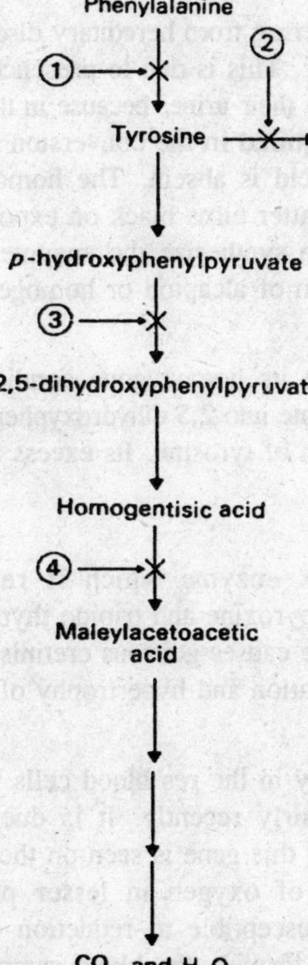

Fig. 13.1. Part of the biochemical pathway for the metabolism of phenylalanine and tyrosine. The sites of metabolic blocks associated with human diseases are indicated by crosses: (1) phenylketonuria; (2) albinism; (3) tyrosinosis; (4) alkaptonuria.

acid. As a result phenylpyruvic acid accumulates in the blood and causes impairment of brain tissue, pale skin and a tendency to epileptic seizure. Thus PKU child-ren are mentally retarded and are known as phenylpyruvic idiots. Such persons have recessive genotype pp that fails to produce enzyme phenyl-alanine hydroxylase.

Persons suffering from phenylketonuria also show mild symptoms of albination; this is because melanin pigment is not produced in such individuals in as much abundance as is necessary for normal colour development.

Alcaptonuria

The urine of persons suffering from hereditary disease alcaptonuria turns black on exposure to air. This is due to presence of a substance adoption (homogentisic acid) in their urine, because in them the enzyme-homogentisic acid oxidase required in the conversion of homogentisic acid into maleylacetoacetic acid is absent. The homogentisic acid is passed out in the urine, and latter turns black on exposure to air. The recessive mutant gene fails to synthesize the enzyme and blocks the step in the normal degradation of alcapton or homogentisic acid.

Tyrosionsis

The recessive gene, t in its homozygous condition, blocks the conversion of p-hydroxy pyruvate into 2,5 dihydroxyphenyl pyruvic acid. This leads to the accumulation of tyrosine. Its excess is excreted with the urine.

Goitrous Cretinism

Sometimes persons lack enzyme which is required for the conversion of thyrosine into thyroxine and trioido thyronine hormone. Deficiency of thyroid hormone causes goitrous cretinism characterized by physical and mental retardation and hypertrophy of thyroid gland.

Sickel Cell Anemia

It is a type of abnormality in the res blood cells whose genetical basis has been determined fairly recently. It is due to a recessive gens. The immediate effect of this gene is seen on the shape of RMC affecting the transportation of oxygen in lesser proportion. The homozygous stage is more susceptible to reduction of oxygen than heterozygous. In fact, in the former, the blood sample may show a few sickel-shaped cells before any substantial reduction of oxygen.

The homozygous recessive individuals rarely survive to leave any progeny because of secondary (*pleiotropic*) effect of the gene (the arms and legs are elongated and there are other physical deformities). The heterozygous individuals are indistinguishable from the normal except that the shape of their RBC changes consequently or lowered oxygen supply. These heterozygotes have been found to be superior over normal in respect of their immunity to the attack of malarial parasites. When the haemoglobin from normal and sickle cell anemic cells were analyzed, they were seen identical in all the fragments excepting for the number four. From the determination of amino acid sequence, it was seen that the fragment four of the abnormal haemoglobin contained valine in place of glutamic acid, other amino acids being identical. In other words, valine was seen to replace glutamic acid and the rest of the 285 amino acids remained undisturbed.

Genetic Control of Drosophila eye Pigments

Garrod's pioneering work provided the first clue to the link between Mendelian factors and proteins. In 1935 *G. Beadle* and *B. Ephrussi* obtained more evidence for gene enzyme relationships in a biochemical pathway in their studies of the synthesis of the eye pigments in the fly, *Drosophila melanogaster*.

There are two pigments in the eye of *Drosophila* the bright red pterins and the brown ommochromes. As we now know, these two pigments are made in two multistep biochemical pathway, each reaction of which is catalyzed by an enzyme. The bright red and brown pigments that are the end products of these pathways become attached to protein granules and are deposited in the eye cells. The combination of the two pigments results in the dull red eye colour characteristic of the wild type. Further, in *Drosophila*, the larval stages contain groups of cells called *discs*, each of which develops into a particular adult structure during metamorphosis. Two discs can be identified as the progenitors of eyes. In their initial experiments, Beadle and Ephrussi showed that an eye disc from a larva can be transplanted into the abdomen of a second larva and disc will develop into an identifiable eye structure that can be found in the abdomen of the adult following metamorphosis. This paved the way for a series of elegant experiments in which eye discs were transplanted between larvae of different genetic constitutions.

At the time if *Beadle* and *Ephrussi's* experiments, three gene lock were known that were implicated in the production of brown pigment; scarlet, cinnabar and vermilion. Mutations in any one these genes resulted in a bright orange eye (the shade of orange differs in each mutant type). *Beadle* and *Ephurssi* first addressed the question of whether embryonic eye tissue from the three mutant types *st* (scarlet), *cn*

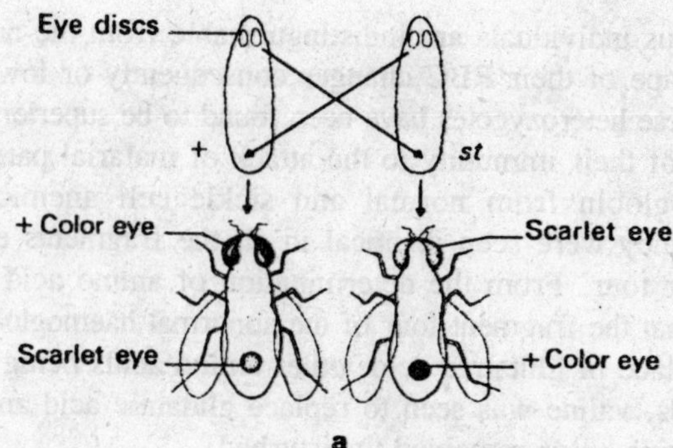

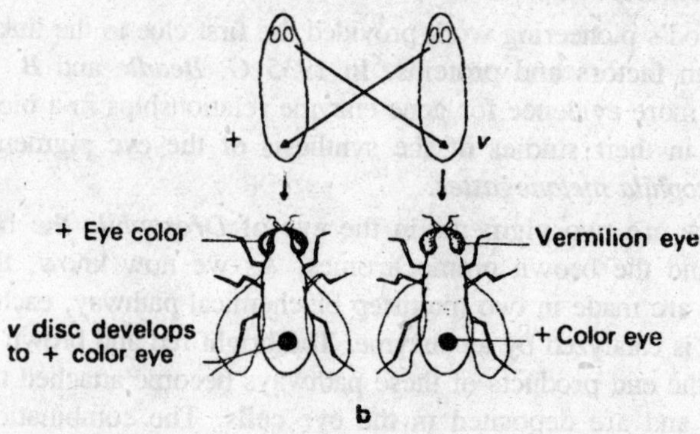

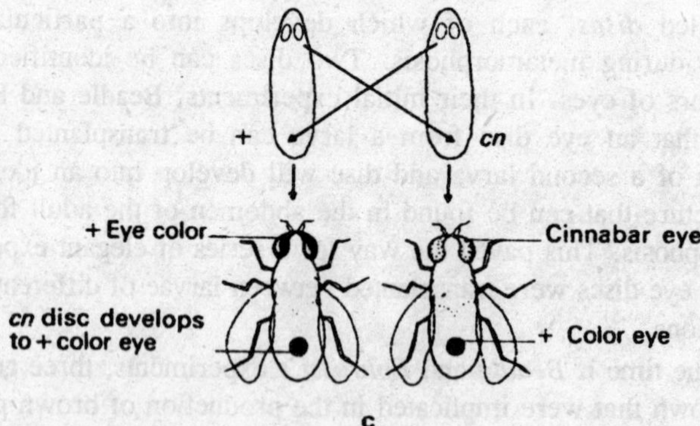

Fig. 13.2. Results of Beadle and Ephrussi's reciprocal transplant experiments with eye discs of Drosophila. (a) Reciprocal transplants between + and st (scarlet). (b) Reciprocal transplants between + and v (vermilion). (c) Reciprocal transplant between + and cn (cinnabar).

(cinnabar) or *v* (vermillion) transplanted into a wild type larval host would develop a wild type eye colour. In the case of the st mutant the transplanted disc developed into a scarlet-coloured eye and s an example of *autonomous development* in which the wild type host was unable to provide substances to the disc to enable the brown pigment to be produced. By contrast, transplanted discs from either *v* or *cn* developed into normal-coloured eyes in the wild type host and thus showed non-autonomous development. The conclusion in these cases was that the wild type provided a diffusible substance that the discs used to bypass the genetic block and to make brown pigment.

Of more importance historically were the result of reciprocal transplant experiments between *v* and *cn* larvae. The experiments showed that *cn* discs transplanted into v hosts produced cinnabar-coloured, eyes, whereas v discs transplanted into *cn* hosts developed into wild type coloured type. They concluded from these results that the production of the brown pigment involved a biochemical sequence with at least two precursors, the wild type having both, *cn* having one, and *v* having neither. Thus, the production of a wild type coloured eye when a *v* diffusible substance so that it becomes wild type. In the reciprocal experiment, the cn disc cannot convert the *v+* substance to the *cn+* substance and the *v* host is deficient in the production of *v+* substance so no brown pigment can be produced.

At the time of these experiments, the relationships between genes and enzymes was known. *Beadle* and *Ephrussi's* work was important historically since it indicated a strong link between a phenotype (in this case eye colour) and genotype (*v* and *cn*) *E. Tatum* further investigated this system by making extracts of v and *cn* flies of homogenization and by injecting the extracts into host larvae. He found that the same results were obtained as for the eye disc transplantation experiments; the injection of a *cn* extract into v larvae resulted in the production of wild-type coloured eyes. This experiment was the first in a series designed to discover the nature of the *v+* and *cn+* substances. At that time, it was difficult to analyze the extracts he had prepared, and his limited analysis showed that the substances were water soluble and low molecular weight. He proposed that the substances were discs is implanted into a *cn* host follows logically. This is the *v* disc cannot make the *v+* substance (the product of the wild type allele of v) whereas the *cn* host can. This *v+* substance diffuses to the *v* discs where it converted to the brown pigment. In other words, the *cn* host makes up for the deficiency of the *v* disc by carries the wild type allele of *cn*, and this substance is then converted to the brown pigment. In other words, the *cn* host make sup for the deficiency of the *v* disc by supplying it with a relatives of amino acids and set about trying to supplement the fly food in a defined way to try to bypass the genetic blocks in the growth pigment pathway and therefore to identify the biochemical precursors in the pathway and therefore to identify the biochemical precursors in the pathway in this way. Several complex media such as peptone resulted in brown pigment formation in adult eyes when fed to *v* and *cn* larvae. One exception was gelatin, which is deficient in tryptophan and tyrosine. Tatum then injected tryptophan into *v* and *cn* larvae and no brown pigment was produced in most experiments. However, in one experiment, brown pigment was produced, and it turned out that this occurred as a result of a bacterial contamination, which resulted in the metabolism of tryptophan into chemical that are required by *v* and *cn* for the production of brown pigment. With the implication of tryptophan metabolism in the production of brown pigment, it was then a relatively simple matter to work out the biochemical pathway and to localize the steps catalyzed by the *v*+ and *cn*+ gene products. As can be seen the *v* mutants are unable to convert tryptophan to formylkynurenine and the *cn* mutants cannot convert formylkynurenine to 3-hydroxykynurenine because the enzymes they have for these respective steps are non-functional.

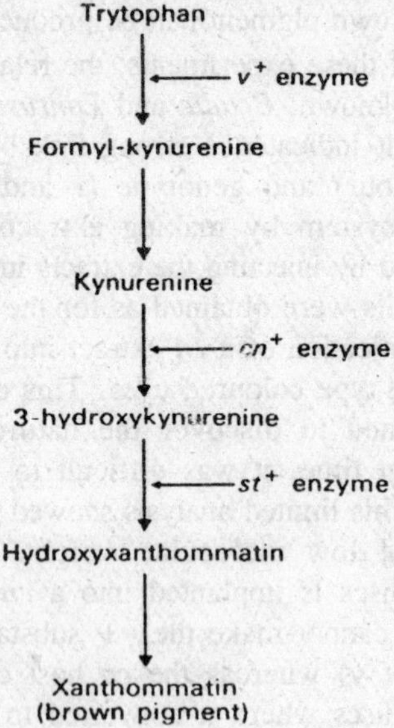

Fig. 13.3. The pathway for the production of brown eye pigment of Drosophila from the amino acid tryptophan.

The pathway can serve to illustrate a general point about the sequence of reactions and the consequences of genetic mutations. In the *v* mutations, for example, the flies accumulate high amounts of tryptophan since it is not converted to the brown pigment in this pathway. This is analogous to the accumulation of homogentisic acid in alkaptonuriacs. Here it provided further evidence that tryptophan metabolism is involved with eye pigment formation. Indeed, if one feeds wild-type larvae with radioactive tryptophan, the brown ommochrome pigment of the adult eye is found to be radioactive.

Biochemical Mutations in Neurospora

The unicellular fungus Neurospora has been most widely utilized in this study of biochemical mutations influencing or amino metabolic pathways, concerned with the synthesis of acids, molecules of nucleic acids and some of the vitamins.

The normal strain of Neurospora is able to grow in the minimal medium containing sucrose, inorganic salts and one of the vitamins–biotin. It can synthesize all other complex substances, like vitamins, proteins, nitrogenous bases and the nucleic acid from the simple substances of culture medium. Beadle and Tatum exposed some of the asexual spores (conidia) of Neurospora crassa to mutagen (X-rays or ultraviolet rays). These mutated ascospores were found unable to grow in minimal medium but required the addition of some supplement for the normal growth. The mutated strains of Neurospora are called *nutritional mutants* showing mutations in the nutritional genes.

Controlled experiments by *Beadle* and *Tatum* have shown that biosynthesis of each amino acid involves a number of sequential steps and each step is regulated by a specific enzymes. The end product of one reaction acts as a precursor for the next one. For example, synthesis of amino acid arginine involves the following steps.

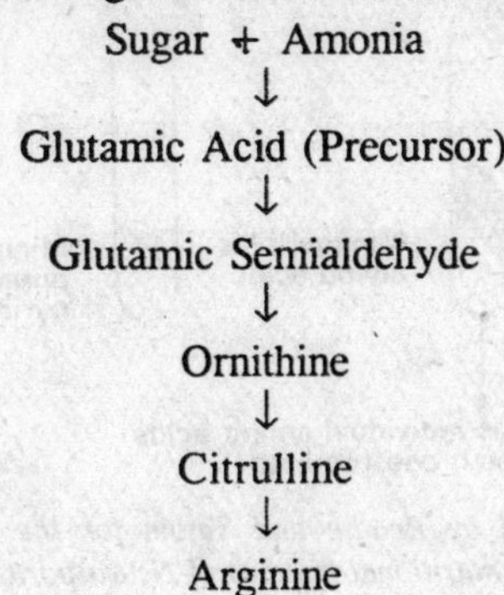

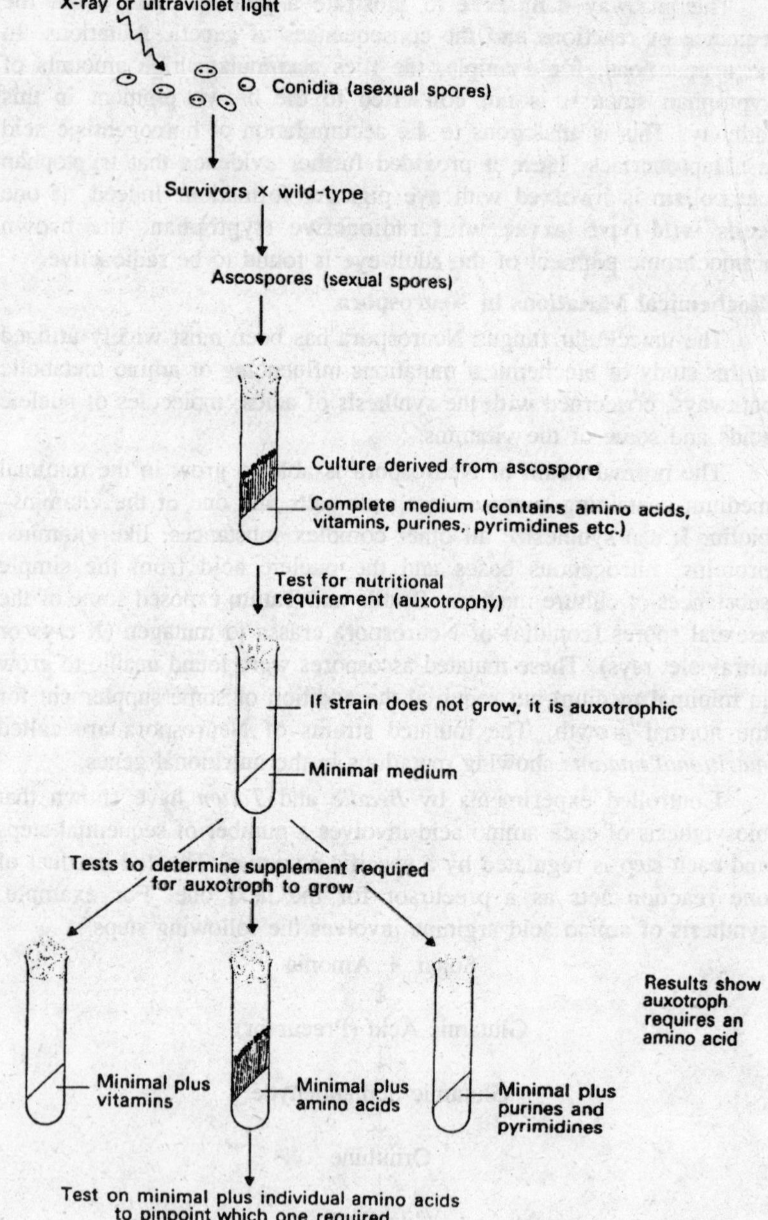

Fig. 13.4. Procedure used by Beadle and Tatum for the induction, isolation, and classification of nutritional mutants of Neurospora.

There are 12 different strains of *N. crassa* which are unable to synthesize arginine. These form three mutant categories.

(a) Mutants which grow only when either citrulline or arginine is supplied but not grow if ornithine or cutrulline.

(b) Mutants which grow when either citrulline or arginine is supplied but not grow if ornithrine is added to the medium.

(c) Mutants which can grow in the medium containing either ornithine, citrulline or arginine.

These mutants indicate that in first case (a) the mutants are incapable of utilizing intermediate products ornithine or citrulline. In them the gene responsible for the conversion of citrulline into arginine has undergone mutation and fails to synthesize the specific enzyme. (b) The mutants are incapable of utilizing ornithine. It means the enzyme responsible for transformation of ornithine into citrulline is lacking and the gene concerned with the synthesis of this enzyme has undergone mutation. In third case (c) the mutant can utilize intermediate products but not the precursor (i.e. glutamic acid).

Biochemical Mutants in Flowering Plants

Arabidopsis thaliana is a small rapidly growing crucifer with about a 28 day life cycle. It has a low chromosome number (2n=10). The number of seeds per fruit is quite large (150-300 seeds) and 25000 per plant. It cannot however tolerate a temperature higher than 25. Requirement of such low temperature therefore makes it extremely difficult to work with this material in the tropics except in winter. When grown under controlled condition, it is supplied with 20 hours photoperiod and high intensities of 10-12 thousand lux. Because of their small size and fast growth rate they can be grown continuously through their life cycle in the test tube on sterile nutrient medium +2 percent sucrose modified in suitable concentration of agar. The trace elements used in the culture solutionis chelated.

By means of various mutagenic agents e.g. X-rays EMS (1 per cent for two hours) etc. auxotrophs are easily obtained. Different types of lethal embryos (vanda, fusca, parwa, diffusa) can be picked up under the dissection microscope. The unripe fruits are opened and the undeveloped embryos can be seen without dissecting the seeds. Chlorophyll mutants appear in the form of albino, X antha, chlorine (yellow green). As in the case of biochemical mutants in *N. crassa*, supply of different growth requiring substances (amino-acids, vitamins) either individually as in combination in the culture medium enables the mutants grow to different lends of development. Some chlorophyll mutants die at the cotyledon stage some at the 3 rosettle stage without turning green, still others at the bud stage. Some of the mutants which deserve special mention are leucine, tryptophan, thiamine anxotrophs. These, on supply of their specific requirements, turn green and grow to full length and produce seeds. One of the interesting observation in these studies 15 that in leucine auxotroph, addition of this amino acid by itself restores the normal growth but in combination with valine in different proportions inhibits growth.

Gene Control of Protein Structure

So far we have learned that there is a direct relationship between genes and enzymes. Since an enzyme catalyzes one reaction in a biochemical pathway, a mutation in the gene for the organism depends on the nature of the pathway, the effects of the compound accumulated prior to the block, and the effects of a deficiency of the end product.

However, while all enzymes are proteins, not all proteins are enzymes. Nevertheless, the functions of proteins are also affected by changes in amino acid sequence and tertiary structure. For example, the human disease sickel cell anemia is characterized by a sickling of the red blood cells. This leads to problems with the red blood cells passing through the capillaries, hereby causing tissue and organ damage, anemia, and possibly death. The life of the red cells is also greatly shortened. Sickle cell anemia is caused by homozygosity for a mutation in the gene for the b-polypeptide of haemoglobin. (Haemoglobin consists of two copies each of a and b polypeptides). The mutation results in the substitution of the neutral amino acid valine for the normal acidic amino acid glutamic acid at the sixth position form the N-terminal end. This change is at an important part of the molecule, and the presence of a neutral amino acid instead of an acidic amino acid makes the region hydrophobic (water-hating) instead of hydrophilic (water-loving). As a result, the b-polypetide folds to place the altered region away from the aqueous environment, and this causes an unusual stacking of haemoglobin molecules and sickling of the red blood cells.

Colinearity

A more precise understanding of the relationships between a gene (cistron) and the sequence of amino acids in the polypeptide for which it codes came from the work of *C. Yanofsky* and his group in 1967. Their studies centered on the enzyme tryptophan synthetase from *E. coli*. This enzyme consists of two copies each of two distinct polypeptides A and B, which are coded for by two adjacent genes. Here is an example of where two genes code for one enzyme—a clear

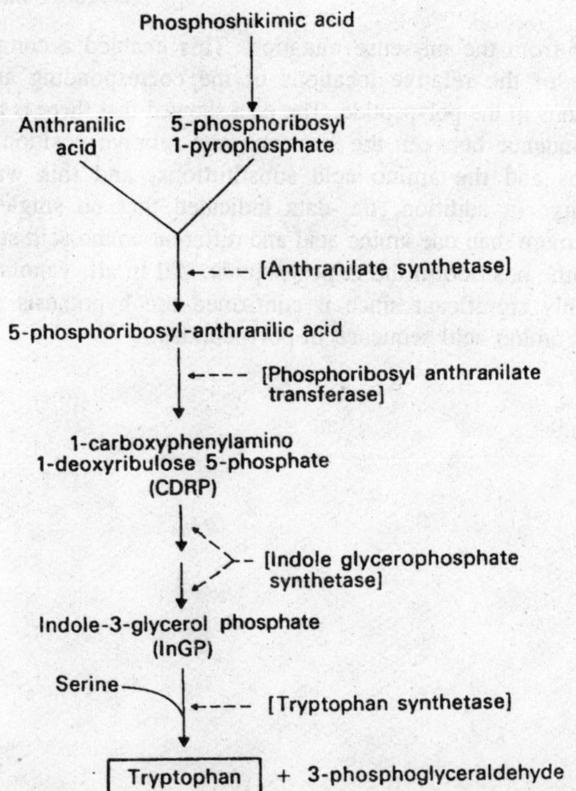

Fig. 13.5. The tryptophan biosynthetic pathway. The enzymes that catalyze the reaction steps are shown in square brackets. Tryptophan synthetase catalyzes the last step in the pathway.

exception to Beadle and Tatum's hypothesis. The reactions that the enzyme carries out are in the biosynthetic pathway for the amino acid tryptophan.

Tryptophan synthetase is easily isolated and the two polypeptide A and B can be purified. Further, the amino acid sequence of the 267-amino acids-long. A polypeptide was determined at the outset of their experiments. They then isolated a series of tryptophan auxotrophs and, with further tests, identified those that carried missense mutations in the tip A gene, which codes for the A polypeptide. The tip A mutants cannot carry out the production of tryptophan form indole glycerol phosphate and serine. Fine-structure mapping was used to locate the various mutations within the gene and amino acid sequencing was used to pinpoint the amino acid substitutions in the A polypeptide resulting from the missense mutation. This enabled a comparison to be made of the relative locations of the corresponding amino acid substitutions in the polypeptide. The data showed that there is a complete correspondence between the sequence and relative positioning of the mutations and the amino acid substitutions, and this was termed colinearity. In addition, the data indicated that no single mutation affected more than one amino acid and different amino acid substitutions at the same position in the A polypeptide. All in all, yanofsky's work was highly significant since it confirmed the hypothesis that genes code for amino acid sequence in polypeptides.

14

ENVIRONMENT AND ITS INHABITANTS

The English word "ecology" is taken from the Greek *oikos*, meaning house, the immediate human environment. In 1870, the German zoologist Ernst Haeckel first gave the word its broader meaning, the study of the natural environment and of the relations of organisms to each other and to their surroundings. General use of the word came only in the late 1800s, when European and American scientists began to call themselves ecologists. The first societies and journals explicity devoted to ecology appeared in the early decades of this century. Since that time, ecology has undergone immense growth and diversification, so much so that persons devoting their professional lives to ecology now number in the tens of thousands. With the dual crises of rapid growth of human population and accelerating deterioration of the earth's environment, ecology has taken on the utmost importance to everyone. Management of biotic resources in a way that sustains a reasonable quality of human life depends upon wise ecological principles, not merely to solve or prevent environmental problems but to inform our economic, political, and social thought and practice.

Nitrogen cycles in forests, energy budgets of pond organisms, stability of marine food chains, pollination systems of plants, mating calls of toads, predator-prey oscillations—all fall within the realm of ecology. This diversity is well illustrated by the complexities that meet our effort toward control and management of populations, whether to maintain resources of food or fuel, to reduce the incidence of disease, or to preserve species of aesthetic or cultural worth. Management of populations requires us to understand the causes of population change and the consequences of population change in any natural system. Several years ago, the Nile perch was introduced into Lake Victoria in East Africa. This was done with the well-intentioned purpose of providing additional food for the peoples living in the area and additional income from export of the surplus catch. Because simple ecological principles were ignored, the result was virtual destruction of the lake's entire fishery. Until the introduction of the Nile perch, Lake Victoria supported endemic fishing for a variety of species of cichlids, which themselves fed primarily on detritus and plants. The large Nile perch feeds upon other fish, the smaller cichlids in this instance. Because energy is lost with each step in the food chain, a predatory fish cannot be harvested at as high a rate as its herbivorous prey. Furthermore, the perch was alien to Lake Victoria, and evolution of the local cichlids did not include behaviour to escape predation. Inevitably, the perch annihilated the cichlid populations, destroying the native fishery and all but eliminating its own food, thereby, in turn, hastening its own demise as an exploitable fish. To be sure, the native fishery was already precariously overexploited owing to growth of the local human population and recent application of less primitive fishing technology, but the appropriate solution to these problems would have been better management of the cichlids, not introduction of an efficient predator upon them.

Still other results of the introduction of the alien fish turn the story into a tragic comedy of errors. The flesh of the Nile perch is not particularly liked by the peoples living near the shores of Lake Victoria. They prefer the accustomed texture and flavor of the native fishes. Moreover, the oily meat of the Nile perch must be preserved by smoking rather than sun-drying, and local forests are being cut rapidly for firewood. Because the larger Nile perch requires larger and more elaborate nets, local subsistence fishermen cannot compete with more prosperous outsiders who are equipped for commercial fishing. The lesson is painful yet simple: Man is an integral part of the ecology of the Lake Victoria area. Traditional local fishing had been sustained for thousands of years until the pressure of population and the perception of an opportunity for an export fishery led to an ecologically unsound decision and to an economic and social disaster.

Half a world away, efforts to save the California sea otter illustrate the intricate intermingling of ecology and other human concerns. The sea otter was once widely distributed around the North Pacific Rim from Japan to Baja California, but in the 1700s and 1800s intense hunting for fine otter pelts reduced the population to near extinction. The fur hunting industry, of Course, collapsed from the overexploitation. Subsequent protection enabled the California sea otter population to rise to more than 1500 individuals by the 1980s. The sea otter's success, alas, has irked California fishermen, who claim that the otters —which do not need commercial fishing licenses—drastically reduce stocks of valuable abalone, sea urchins, and spiny lobsters. The matter has deteriorated to the marine equivalent of a range war between fishermen and conservationists, with the otter, often fatally, caught in the line of fire. Ironically, the otters benefit a different commercial marine enterprise, the harvesting of kelp, a large seaweed used in making fertilizer. The kelp grows in shallow waters in stands called kelp forests, which provide refuge and feeding grounds for several larval fish. The kelp is also grazed by sea urchins which, like goats on land, can, given the right conditions, denude an area of vegetation. One of those conditions is the absence of a principal sea urchin predator, the sea otter. When the expanding sea otter population has spread into new areas, urchin populations have been controlled, allowing kelp forests to regrow. Human involvement in the local ecosystem requires that management of otter populations and various commercial exploitation be balanced in an ecónomically and socially acceptable manner, an impossibility without understanding the complex ecological role of the otter in the system.

Although the plight of endangered species may arouse us emotionally, there is a beginning realization that the only means of preserving and using natural resources is through the conservation of entire ecological systems and of broad-scale ecological processes. Individual species, including the ones that humans rely upon for food and other products, themselves depend upon the maintenance of a healthy ecological support system. We have already seen how such predators as the Nile perch and sea otter may assume key roles in maintaining the natural properties of a system or in dramatically altering its properties. On land, pollinators and seed-dispersers are necessary to the long-term maintenance of plant formations. Therefore, the management of forests without attention to maintaining populations of insects, birds, and mammals, upon which the long life of the forest depends, is ecologically naive. Natural systems are diverse and highly evolved in the sense that each species is adapted to its particular role within the system and the system depends to some degree on each player acting out its role. It turns out that any reduction in biodiversity can upset the balance of a system and alter its function.

Most natural systems sustain cycles of disturbance (fire, hurricanes, disease) and rejuvenation. Natural restoration depends on there being patches of intact habitat from which species may recolonize disturbed areas. But as habitats become more and more fragmented, as are forests by the spread of agriculture and urban development, disturbances can so thoroughly destroy an area as to leave little chance of complete recovery, even with substantial help from man. Disturbance cycles, whatever their spatial and temporal scale, are as much a part of the natural landscape as are key species within a local system and must be taken equally into account in any effort to maintain the system.

Many of the now rapid changes in the world bring a seeming improvement in the quality of human life, but those very changes increasingly strain the earth's ecological support systems and in reality bring decline in the quality of life. On any scale on which we view the surface of the earth, the deterioration of the environment is alarming. Partially enclosed systems, such as the estuaries of major rivers, allow us to isolate the causes of much of the change. In San Francisco Bay, for example, the pressures of population and agriculture have gradually but profoundly altered it to the point that it would be unrecognizable to nineteenth-century inhibitants of the region. Diking of wetlands to create farmland, filling to create land for urban construction, pollution from sewage and from agricultural and industrial waste, and introduction of exotic species have transformed a healthy, productive, and economically valuable system into an ecological disaster, whose greatly diminished value lies entirely in recreation, itself of ever less value, and in commercial shipping, not in its once generous biotic resources. The present human population of the San Francisco Bay area may well be incompatible with a pristine estuary, but much of the change came about through state subsidy of unprofitable agricultural ventures, absence of ecologically and economically sensible city planning, careless introduction of pest species, use of the bay as a cheap sewer, and uncontrolled harvesting of fish and shellfish.

The multiplication of technology and population has brought ecological crisis to far larger regions of the earth than San Francisco Bay. One of the most critical is the Sahel in Africa. The Sahel is a vast band of semiarid habitat lying at the southern edge of the Saharan Desert and extending from Gambia in the west to Ethiopia in the east. Population pressure, expressed through intensified grazing and collection of firewood, has caused extensive deforestation of the entire region. As humans and domesticated animals clear vegetation over such vast areas, climate is affected. Rainfall decreases because less water vapour is recirculated to the atmosphere and heating and cooling patterns of the land surface are changed. Even when plentiful rains result in good yields from crops, the abundance may be snatched away by plagues of grasshoppers and locusts, whose populations grow rapidly under the moister conditions. The Sahel is at present truly devastated.

In wetter parts of the world, tropical rainforests are being cleared at an alarming rate for forest products and largely unsuitable agriculture, bringing profit to a few in the short term and yet another disaster in the long term. The destruction of rainforests dismantles an intricate, self-sustaining system that cannot be rebuilt. Species, most of them unknown to science and many of them with potential commercial value, are being driven to extinction. Burning of cut vegetation adds greatly to the already dangerous level of carbon dioxide in the atmosphere. In this instance, the ecological principles that should govern human use of the rainforest are generally understood. 'As Michael Robinson, director of the National Zoo, Washington, D.C., has pointed out, "The problems are not (the result of) ignorance and stupidity. The problems derive from the poverty of the poor and the greed of the rich".

The destruction of rainforests is scarcely alone in its far-reaching effect. Automobile emissions reduce tree and crop production over large areas. Sulfurous gases from burning coal have acidified the rain and snow over large areas of eastern North America and Europe, killing forests and degrading freshwater ecosystems. Chlorofluorocarbons (CFCs) used to pressurize spray cans and as refrigerants have reduced the ozone concentration in the upper atmosphere, increasing the amount of dangerous ultraviolet radiation that reaches the earth's surface.

As a result of industrial growth and deforestation, levels of carbon dioxide, methane, and CFCs have increased in the atmosphere so much that they are changing the earth's climate. In the so-called greenhouse effect, long-wave radiation from the earth's surface is absorbed and retained by the atmosphere. As a result the temperature of the earth is increasing and is expected to rise by 2 to 6 degrees Celsius during the next century. Such change will cause a rise in sea level as the polar ice caps melt, it will disrupt agriculture and shift ecological habitats across the landscape as temperature and moisture patterns change, and it will perhaps cause widespread extinction of forms of life unable to adjust to the change and its effect.

Our ability to wreak havoc upon ourselves and the world in which we live is virtually unlimited. The most appalling evidence of this ability is our means of waging global nuclear war, wherein destruction of ecological systems and extinction of species, possibly including ourselves, would be virtually without parallel in the history of the earth. Nuclear war remains unthinkable even if we ignore the ecological consequencies. What we are gradually accomplishing in the degradation of a livable planet may end by being just as punishing. Merely to understand ecology will not solve our environmental problems in all their political, economic, and social dimensions. But as we contemplate the need for global management of natural systems, we are doomed to fail if we do not understand their structure and function, an understanding that depends upon the principles of ecology.

15

Improving Human Race

The science of eugenics was founded in 1883 by *Sir Francis Galton*, a cousin of *Charles Darwin*. The idea itself is very ancient and can be found in the works of *Plato* and the sages of antiquity. It as much as many diseases, defects, and abnormalities are hereditary and some are regularly transmitted by parents to children, it was not unreasonable to propose that children unfit for life should not be brought into the world. This, of course, amounts to preventing pèrsons who are seriously defective from reproducing.

Legislation to sterilize such persons exists in many countries in the United States the first law of this kind was adopted in 1907 by the State of Indiana. By 1940 similar laws existed in thirty states, being more strict in some than in others. In all thirty states mental deficiency is subject to sterilization; in twenty-nine mental diseases are liable to the same treatment. In twenty-three states there are laws controlling epilepsy. In other states sterilization can be applied in certain cases of physical malformation and nervous diseases. From 1907 to 1950 more than 50,000 such operations were performed, two fifths of the sterilized being men. It is still difficult to weigh the results of these eugenic experiments which have not in all cases been based on sufficiently solid genetic and medical grounds. Similar legislation has been proposed in certain provinces of Canada, in Tasmania, and in New Zealand.

In Europe sterilization laws were first introduced in 1929 in the Swiss canton of Vaud, and in the same year in Denmark, Norway, Sweden, Finland, and Iceland. In Germany and in Estonia the law, after being blindly applied, has been repealed. Laws have been proposed, without being passed, in England, Holland, Hungary, Czechoslovakia and Poland. At the moment such laws are retained only in the Scandinavian countries and in some Swiss cantons.

The Scandinavian laws provide that the patient himself can ask to be sterilized and that he operation cannot be performed without his freely given consent. During the period 1929–1950 more than 6,000 operations took place in Denmark, two-thirds of those sterilized suffering from mental deficiency, and two-thirds of these being women. In Sweden, for the period 1935–1948, some 15,650 people were sterilized, women again outnumbering men. The number of those sterilized continues to rise and Professor *Kemp* estimates that in Denmark today there are 600 operations a year, distributed equally between the mentally deficient and those suffering from other afflictions. At this rate an increasingly rapid elimination of mental deficiency may be anticipated. Denmark not only has laws providing for sterilization but also others covering legal abortion, the prevention of certain marriages, and eugenic measures such as isolation, the obligatory declaration of disease etc.

Of course, these various measures encounter moral, social, and religious opposition, and the question arises whether their efficacy is genuine. The answer will vary according to the way in which the gene responsible for the affliction is transmitted.

Eugenics can be studied under two heads:

(i) Negative eugenics.

(ii) Positive eugenics.

Negative Engines

In this the defective germ plasm is not allowed to mix. This restriction can be brought about in the following ways:

By Controlling Immigrations

Most of the advanced nations are very selective in permitting foreigners to enter their lands and mingle with their populations. They believe in encouraging only the *cream* of other nations, i.e. people with high intellect. This is an attempt to shut out germplasm which may be undesirable.

It is compulsory to get vaccinated and injected as an immunity measure, against certain contagious and epidemic diseases before travelling to a foreign country.

But these measures are not sufficient to check undesirable germplasm because a person may have high intellect and may possess sound health but in his genome he may be carrying an undetected recessive gene which when homozygous may be fatal.

Marriage Restrictions

Every country or society has the customs or laws which tend to restrict marriages. According to biological aspects marriage is an experiment in breeding, not human courtship as described by various novels. In several countries the marriages of mental defectives, habitual drunkards, idiots, feeble-minded, insane persons, epileptics, alcoholics and persons having venereal diseases are prohibited. As a result of these regulations the genes necessary for these defective traits do not have chance to mix in the generation. Only marriages between undefective persons should occur. Even the cousin marriages are extremely useful if they are without set-linked diseases.

Sexual Separation of the Defective

The defective persons may have various sex-linked diseases such as night blindness, haemophilia, colour blindness, etc., and various other defective traits which may be regulated by dominant or recessive genes. The increase of germplasm of the persons having such defective traits in the population can be checked by keeping them away and separated from the society. Different states have wisely adopted the restricted measures in segregation the mental defectives from the society and to place them in mental hospitals.

Sterilization

It is a drastic step of restricting defective germplasm from meeting with the other germplasm. By this method, ducts carrying sperms or eggs from gonads are removed surgically thus blocking the way of undesirable germplasm to external world. This is of two types :

Vasectomy

It is minor operation in males in which vasa deferentia (sperm ducts) are surgically removed or rendered ineffective by some means. In the castration complete testes are removed thus putting the sexual life to an end.

Salpingectomy

In this type oviducts are gutted or removed surgically in females and thus avoid the chance of fertilization due to non-availability of ovum.

Thus, these operations fall into two categories namely temporary. and permanent. In the temporary sterilization, vasa deferential or fallopian tubes are tied-off with the help of surgical thread in males or females. In the latter method, these ducts are cutted permanently.

Criteria for Negative Eugenics (Genetic Variables)

In practice the effectiveness of marriage restrictions, segregation, control on immigrations and sterilization as eugenics measures depends on several genetic variables like:

(a) The mode in which undesirable trait is inherited;
(b) The rate of frequency of the gene or genes concerned in the populations;
(c) The age at which the defective trait makes its appearance; and
(d) The extent to which environment may check or inhibit the expression of the gene. Regarding the mode of inheritance of these trails the following possibilities may exist; (i) The trait may be due to dominant gene. (ii) It may be due to a gene without dominance (i.e. heterozygote intermediate). (iii) Sex-linked recessive gene may bring about it. (iv) Finally, a trait may be produced as a result of cumulative genes.

Positive Eugenics

Subsidizing the Fit and Sperm Banks

Because the highly endowed persons lead a well-planed life and to avoid unnecessary difficulties in nursing the children they often prefer to have small number of children. Therefore, the selected young men and women of best eugenic value should be encouraged to increase their birth rate. Moreover, *H.J. Muller* has suggested that such persons not only should increase their family size but through artificial insemination the outstanding man can serve as fathers to many more children than would be otherwise possible. The artificial insemination is already widely practiced to permit those women whose husbands are sterile or have some serious hereditary afflictions to bear children. The sperms and eggs of outstanding persons can be stored for future use by quick freezing and storing them in deep freeze. These germ cells thus can be stored for 100 or more years.

Recently in U.S.A., a "sperm bank" was floated, primarily for collection and preservation of sperms of Nobel Prize laureates and probably other exceptionally gifted intellectuals. Its aim is to produce children of "high mental quality" through artificial insemination. The initial response to the sperm bank was poor but recently *Robert Graham* has claimed that the response is now fairly good. It is also claimed that about 15 children have already been produced with the help of sperms thus acquired. However, it is yet to be seen how far this revolutionary idea succeeds in its aim.

Promotion of Genetic Research

The mechanism of inheritance of many specific traits can be clearly understood provided basic principles of heredity and distribution of genes are known. The physiological action of genes is another problem and with its complete knowledge, many diseases can be cured. The nature of mutation, their effects and possible control may bring about the better offsprings having high learning capacity.

Besides, genetic research on plants and animals is receiving much support from different state government. By this encouragement, good varieties of plants and animals highly beneficial to man have been produced.

Early Marriage of those having Desirable Traits

It is most commonly observed fact that the highly placed persons of the society often have great ambitions for the future life. In achieve their ambitions goals, they often devote the best part of their youth and they are able to marry in their mature egg (i.e., 30 to 35 years). The biological and psychological investigations have revealed that the aged persons often lack in necessary amount of emotional warmth for the sexual activities and moreover, their germplasm also lost its vigour. Therefore, some laws should be formulated to prevent the late marriages of highly endowed persons by applying high taxation on them and at the same time the young persons having best hereditary traits should encouraged for early marriages.

Genetic Counselling

To produce healthy progeny should be the endeavour of man as this alone can ensure a better future for humanity. Genetic counselling can make a significant contribution in this direction. For families with history of genetic diseases, genetic counselling can provide the much needed relief.

By careful examination the genetic counseller can detect genetic abnormalities like *Down's syndrome* and even recessive genes for disease like *sickl-cell anaemia*. Knowledge of the possibilities of transmitting these diseases to offspring can help a person in choosing a marriage partner.

Genetic counselling is equally useful after marriage. An Rh^- woman with an Rh^+ partner, when aware of the implications in advance, can go in for suitable medical aid well in time. The study of the amniotic fluid (obtained by a technique called *amniocentesis* involving withdrawal of the fluid with the help of a hollow needle) can reveal the sex of the foetus as well as *congenital diseases* (like Down's syndrome) and metabolic disorders that it may possess. This information can help the couple to decide whether or not retain the pregnancy, or at least be mentally prepared for the inevitable.

Protection against Mutagens

We now know that mutagens can induce mutations, some of which may be deleterious and even lethal. Harmful mutations can result in deformed offspring and increase human misery. It is, therefore, essential to check exposure of human germ cells to mutagens like high-energy irradiations, chemicals, etc. Medical X-ray should be advised and carried-out with utmost care. Chemical mutagens should be experimented, with, and handled with all possible precautions. The December 3, 1984 gas tragedy in Bhopal (M.P.), India, (leakage of MIC–methyl isocyanate) and the subsequent discovery of "dangerous" experiments carried out in Bhopal should be sufficient to warm world governments of the possible hazards of chemical warfare in which chemical mutagens may be used. These hazards will in no way be less then those of exposure to high-energy irradiations frigerred by atomic bombs.

Improvement of Environment Conditions

Environment plays a dominant role in the heredity of man. Better facilities of training, schooling, living and studying should be given. These all are euthenic programs for the benevolence of human race.

Generally sexual fertility is correlated with intellectuality reaching maximum very rapidly in idiots and stupids while more slowly in intelligent and genius persons. A shown earlier idiots as well genius are homozygous genes responsible for particular traits while average are heterozygous having greater fertility. Their selection may yield good eugenic results.

EUPHENICS

Euphenics deals with the treatment of genetic diseases of man. The credit for the goes to A.C. Pai (1974). More specifically it deals with the control of several inherited human diseases especially in born errors of metabolism in which the missing of defective enzyme has been identified. The excellent example of two category is the *phenylketonuria* or PKU. In the abnormally the patients are unable is metabolize an amino acid–the *phenylalanine* properly, the resulting product causes severe mental retardation. Such patients are advised to take phenylalanine free diet.

Although a number of inherited diseases can be treated in a similar euphenic manner, but these constitute only a small fraction of known inherited diseases. For the most part, biochemical geneticists could not identified the biochemical errors of many genetic diseases. In other cases, such as *albinism*, even though the metabolic block leading to an abnormality is known, but, it is not possible to correct it.

INDEX